Springer Climate

Series Editor

John Dodson, Institute of Earth Environment, Chinese Academy of Sciences, Xian, Shaanxi, China

Springer Climate is an interdisciplinary book series dedicated to climate research. This includes climatology, climate change impacts, climate change management, climate change policy, regional climate studies, climate monitoring and modeling, palaeoclimatology etc. The series publishes high quality research for scientists, researchers, students and policy makers. An author/editor questionnaire, instructions for authors and a book proposal form can be obtained from the Publishing Editor.

Now indexed in Scopus® !

Kholoud Kahime · Mohamed El Yamani ·
Stéphane Pouffary
Editors

Climate Change Effects and Sustainability Needs

The Case of Morocco

Springer

Editors
Kholoud Kahime
School of Technology Essaouira
Cadi Ayyad University of Marrakesh;
CI2RC
Essaouira, Morocco

Mohamed El Yamani
School of Technology Essaouira
Cadi Ayyad University of Marrakesh
Marrakesh, Morocco

Stéphane Pouffary
ENERGIES 2050
Biot, Alpes-Maritimes, France

ISSN 2352-0698 ISSN 2352-0701 (electronic)
Springer Climate
ISBN 978-3-031-59602-5 ISBN 978-3-031-59603-2 (eBook)
https://doi.org/10.1007/978-3-031-59603-2

This Springer imprint is published by the registered company Springer Nature Switzerland AG
The registered company address is: Gewerbestrasse 11, 6330 Cham, Switzerland

Paper in this product is recyclable.

Foreword

Morocco's geographical location is both a curse and a blessing. The kingdom's natural borders are the Atlantic Ocean and the Sahara. Both will expand as a result of climate change and either flood or dry up the livelihoods of thousands of Moroccans. Many are already being driven from the countryside to large urban centres such as Casablanca. In 2022, Morocco experienced the longest period of drought in more than 40 years, which resulted in serious crop failures and real price explosions for agricultural products.

Moroccans will continue to feel the effects of climate change in the coming years, particularly in the form of water shortages. According to calculations by the World Bank[1], the amount of sustainably available water per inhabitant has fallen from more than 2500 cubic metres in 1960 to just 600 cubic metres in 2020. This means that Morocco is on the verge of sliding from an unproblematic supply situation to a state of water emergency within half a century.

On the other hand, Morocco has immeasurable potential for renewable energy. This is thanks to the high level of solar radiation, the reliable trade winds in the south and—albeit increasingly less—the water flowing from the Atlas Mountains. According to an estimate[2] by the German Society for International Cooperation (GIZ), Morocco could cover its entire electricity needs with hydro, wind and solar energy in the medium term and at the same time export surplus energy, for example in the form of green hydrogen.

Morocco pursues an ambitious climate and environmental policy. The country already covers almost 20% of its electricity needs with renewable energy. If all installed capacities were utilised, this figure could even rise to almost 40%. By 2030, more than half of the electricity is to be generated by water, wind and sun. This makes the kingdom a leader among developing countries and an important partner

[1] https://www.banquemondiale.org/fr/news/press-release/2022/07/20/moroccan-economy-slows-in-wake-of-drought-and-commodity-price-rises.

[2] https://www.irena.org/-/media/Files/IRENA/Agency/Publication/2021/March/Renewable-Energy-Transition-Africa_Country_Studies_2021.pdf?la=en&hash=46D8ADDF378CD917C90F85F899B3F2B33A787CB8.

in the international fight against climate change. A highly professional academic community and an extraordinary active civil society support this ambition.

This book is the product of the 4th edition of the International Congress on Climate Change (CI2C_2023) on the topic of "Climate, Security and Development: Challenges and Opportunity for a Just Transition" which took place in Essaouira, on 24–26 October 2023, organised by the International Center for Research and Capacity Building (Centre International de Recherche et de Renforcement de Capacités, CI2RC, Essaouira, Morocco), the Friedrich Naumann Foundation for Freedom (FFN, Germany), the ENERGIES 2050 international association (France) and the Higher School of Technology of Essaouira, Cadi Ayyad University (ESTE, UCA, Morocco). This congress gathered researchers from a dozen countries and provided the authors of this book with a platform to present and discuss their findings and hypotheses.

May this book also inspire you and provide you with new insights and perspectives.

Enjoy the read!

February 2024

Sebastian Vagt
Director
Friedrich Naumann Foundation for Freedom Morocco
Rabat, Morocco

Contents

Part I
Climate Observation and Prediction in Morocco

Chapter 1
Exploring Climate Change: Morocco in Focus

Kholoud Kahime, Abdelali Boussakouran, Moulay Abdelmonaim El Hidan, and Mohamed El Yamani

Abstract Manifestations of climate change have been occurring more frequently in recent years, posing major challenges that require that require measures, and even strategies, for adaptation and mitigation. This review starts by providing an overview of climate change, focusing first on defining key concepts and understanding the emergence of associated events beyond natural phenomena. The causes of these changes are then explored, with a heightened emphasis on human-induced factors, including the contribution of industrial activities, agricultural practices, burning of fossil fuel, and greenhouse gas emissions responsible for global warming. Afterwards, the various possible future evolutions of these changes were presented, addressing fundamental principles, scenario construction models, as well as their similarities and differences. This first part concludes by an analysis of the repercussions of such changes on water resources—currently considered the most affected environmental element and will continue to face more impacts—along with water-related sectors and their vulnerability in the face of these challenges. The second part of this review highlights climate change in Morocco, beginning with the country's particular climatic features based on its topography and geographical location. Future

The original version of the chapter has been revised: The corresponding author and the sequence of the authors have been updated. A correction to this chapter can be found at https://doi.org/10.1007/978-3-031-59603-2_13

K. Kahime (✉) · M. El Yamani
Laboratory of Applied Sciences for the Environment and Sustainable Development, Essaouira School of Technology, Cadi Ayyad University, Km 9, Route d'Agadir, B.P. 383, Essaouira Al Jadida, Morocco
e-mail: k.kahime@uca.ma

M. El Yamani
e-mail: elyamani.med90@gmail.com

A. Boussakouran
Laboratory of Natural Resources and Environment, Polydisciplinary Faculty of Taza, Sidi Mohamed Ben Abdellah University, B.P. 1223, Taza-Gare, Taza, Morocco

M. A. El Hidan
Laboratory of Biotechnology and Valorization of Natural Resources, Faculty of Applied Sciences, Ibn Zohr University, Agadir, Morocco

K. Kahime et al. (eds.), *Climate Change Effects and Sustainability Needs*, Springer Climate, https://doi.org/10.1007/978-3-031-59603-2_1

projections are produced on the basis of previous scenarios, mainly for temperature and precipitation, to gauge the extent of climate change at national level. Subsequently, the repercussions of changes that have already occurred, as well as potential impacts resulting from future changes, are examined with a particular focus on the water, agriculture, and food security sectors. In fact, the rise in temperatures, coupled with irregularity and decreased precipitation, have led to a 35% reduction in surface compared to the period of 1940–1970 and a 60-m decline in piezometric levels in certain aquifers. These changes significantly impact water supply, especially for the agriculture sector, which heavily relies (87%) on rainfall conditions, putting food security at risk. Overcoming these critical situations will require substantial efforts, especially as the driving factors are expected to worsen.

Keywords Climate change · Anthropogenic factors · Water resources · Agriculture · Morocco

Introduction

Climate change is currently recognized as a significant global concern for humanity. This is evident in the shifting weather patterns, which play a crucial role in the development and growth (Bazzaz and Sombroek 1996; Hirich et al. 2016).

In Morocco, climate change is a significant challenge, as it is in all Mediterranean countries. Indeed, according to Rochdane et al. (2014), North African countries, including Morocco, are projected to be heavily impacted by the negative consequences of climate change. This includes further decreases in precipitation and increases in temperature, as indicated by Simonneaux et al. (2015) and Tramblay et al. (2014a, b). Its effects include environmental deterioration, reduced agricultural output, and heightened concerns over water, food availability and economic and societal stability in these regions (Ouhamdouch et al. 2018; Trabelsi and Zouari 2019; Bahir et al. 2020; Schilling et al. 2020).

Focusing on Morocco, a rising temperature trend has been well recorded over the past 30 years, exceeding the global average (+ 0.42 °C/decade on average since 1990 for Morocco versus + 0.28 °C/decade for all countries) (Woillez 2019). As for rainfall, El Ajhar et al. (2018) noted a shift over the period 1980–2015 towards drier conditions. These changes have led to disastrous natural phenomena such as drought, flooding, salinization, silting, etc. (Bouaicha and Benabdelfadel 2010). As a result, there is increasing pressure on natural resources, especially those used by the agricultural sector, which is integral to Morocco's economic, social, and environmental development, contributing 13–14% to the country's GDP (Abdelmajid et al. 2021). In relation to water resources, crucial for agricultural production, Morocco is currently classified as one of the countries with high water stress (Mason et al. 2019). In fact, water availability in Morocco has declined from over 2500 m^3 per person in 1960, to 730 m^3 per person in 2005 and 650 m^3 per person in 2015, and is expected to drop below 500 m^3 by 2030 (Mason et al. 2019; Bennouna 2020).

Risks of food insecurity, conflict and population displacement can increase under severe water scarcity and drought conditions. Coastal flooding from sea-level rise, environmental degradation, deforestation and loss of biodiversity and ecosystem services pose serious challenges to current development initiatives, leading to a significant increase in vulnerability to risks and hazards. Hence, there is a growing importance attached to the implementation of sustainable adaptation and resilience measures (USAID 2016).

The objective of this review is to conduct a comprehensive analysis of the evolution, impacts, and challenges associated with climate change, both on a global level and with a specific focus on the Moroccan context. We initiate our examination by exploring fundamental concepts, history, and evolution of climatic events, as well as the diverse factors contributing to climate change. Subsequently, we provide readers with a nuanced understanding of models and future projections of this complex phenomenon. Our review extends beyond a simple presentation of facts and statistics related to climate change, moving towards an in-depth discussion of the resulting impacts on crucial sectors, along with adaptation and mitigation challenges. In the second section, attention turns to the national level, highlighting first meteorological characteristics and their specific future projections. This exploration concludes by examining the repercussions of climate change on the water sector and its related activities, while analyzing their influence on agriculture and food security. Furthermore, we address the complex challenges that arise, requiring adaptation and attenuation strategies to cope with these major changes.

Comprehensive Insights into Global Climate Change

History and Meaning

Climate change refers to a long-term disruption of the Earth's climate and weather patterns. It includes changes in temperature, precipitation, wind patterns and other climatic factors (Kumar et al. 2018). These modifications can be ascribed to both natural processes and human activities. Actually, the Intergovernmental Panel on Climate Change (IPCC) defines climate change in the broadest sense as "any change in climate over time, whether due to natural variability or human actions". However, discussions on climate change in the recent decades have mostly focused on linking it to human-induced alterations. This is especially evident in the rise of greenhouse gas emissions resulting from activities like the combustion of fossil fuels, deforestation, and industrial processes. In fact, the Framework Convention on Climate Change (FCCC) specifically characterizes climate change as "a change of climate which is directly or indirectly attributed to anthropogenic activity, altering the global atmosphere's composition and contributing to natural climate variability over comparable time periods" (Pielke Jr 2004).

Historically, discussion of climate change dates back to the nineteenth century, when Greeks and Americans were worried about the impact of human activity on local climate, especially that of deforestation on rainfall in a specific region. However, the most significant climate changes have occurred naturally. In 1896, evidence emerged that fossil fuel combustion releases carbon dioxide, thereby contributing to global warming, although some critics minimized the human influence on extended climate cycles. By the 1930s, observations had confirmed significant warming in the United States and the North Atlantic region. During the 1950s and 1960s, when government funding increased, a number of scientific studies corroborated carbon dioxide accumulation in the atmosphere as a cause of global warming. In fact, while climate models predicted rising temperatures over the next century, there was no consensus on the need for immediate political actions. At the end of the 1970s, a resumption of global temperature increases was recorded, giving many climatologists the conviction that this trend would continue as greenhouse gases accumulated. Since 1988, experts have determined that without measures to reduce emissions, a doubling of carbon dioxide in the atmosphere would occur by the end of the twenty-first century, leading to a temperature increase of 2–3 °C at the Earth's surface. In 1990, the United Nations Framework Convention on Climate Change (UNFCCC) was adopted following recognition by the international community of the need for coordinated action, giving birth to annual Conferences of the Parties (COPs) designed to discuss climate-related issues. Then, in 1989, the Intergovernmental Panel on Climate change (IPCC) was created under the United Nations auspices to develop a scientific approach of climate change and the related political and economic outcomes. Two years later, IPCC reached a consensus emphasizing the significant risk of global warming. It stressed that future climate change primarily depended on the human policies chosen to manage greenhouse gas emissions. Scientific evidence of climate change has been strengthened, and the IPCC reports (2007) highlighting its effects on ecosystems, sea levels, extreme weather events and human societies (Agrawala 1998; Weart 2008; Sharma et al. 2012).

Causes and Future Projections

Anthropogenic Drivers of Climate Change

Climate change can be driven by both natural and human-induced (anthropogenic) factors that alter the radiation received from the Sun or emitted to space, or that modify the redistribution of energy within the atmosphere as well as between the atmosphere, land, and ocean (USEPA 2017). Human activities that contribute to climate change involve fossil fuel burning, deforestation, agricultural practices, urban expansion and pollutant emissions. These activities release greenhouse gases (GHGs) into the atmosphere, including but not limited to carbon dioxide (CO_2), nitrous oxide (N_2O), methane (CH_4), and water vapor (Meehl et al. 2007). These play a key role in climatic regulation. In fact, without GHGs the average temperature on Earth would be − 18 °C instead of + 14 °C and life might not exist (Omer 2008). However, GHGs from human activities are the main driver of climate change. Since the nineteenth century, humans have significantly increased the level of GHGs in the atmosphere. As a result, the natural climatic balance is being altered and the climate is being readjusted by a warming effect or positive climate forcing. In the period from 1990 to 2015, the worldwide net GHGs emissions generated by human activities had risen by 43% (Oreggioni et al. 2021). Additionally, carbon dioxide emissions accounting for about three-quarters of total emission, increased by 51% during the same period. Moreover, the total warming due to anthropogenic GHGs in Earth's atmosphere has increased by 45% from 1990 to 2019. The global warming effect attributed to carbon dioxide alone had increased by 36%. Population and economic factors are the main drivers of rising CO_2 emissions, with those from fossil fuel combustion and industrial processes accounting for around 78% of the total increase in greenhouse gas emissions (Oreggioni et al. 2021).

As for the global agri-food sector, more than 30% of global energy demand is primarily fulfilled by fossil fuel sources, leading to the emission of approximately 22% of total anthropogenic GHGs (FAO 2011). Burning fossil fuels, cement production, and other activities, elevate CO_2 levels, which reduce the CO_2 absorbed by trees. In addition, human activities related to farming using nitrogen fertilizers, raising livestock, paddy rice cultivation, landfill disposal and natural gas use, increase concentrations of nitrous oxide (N_2O) and methane (CH_4) in the atmosphere, making a substantial contribution to climate change (Fakana 2020).

Forests are well-known for their ability to absorb carbon dioxide from the atmosphere and convert it via photosynthesis into biomass, thereby acting as CO_2 sinks in a natural state (Fakana 2020). This process significantly contributes to the regulation of air and surface temperatures (Haghipour and Burg 2014; Güçlü 2014; IPCC 2019). The various forms of deforestation release the accumulated carbon dioxide stock into the atmosphere (Tahir et al. 2011; Ellison 2018). Periods of catastrophic rainfall followed by prolonged periods of drought have resulted from the disappearance of forest cover (Strasser et al. 2014).

Many other industrial and agricultural processes emit pollutants (other than GHGs) that generate aerosols including black carbon particles, which can induce either a warming or cooling effect (Melillo et al. 2014; Sims et al. 2015; IPCC 2019). Furthermore, fluorinated gases such as chlorofluorocarbons (CFCs), hydrochlorofluorocarbons (HCFCs), perfluorocarbons (PFCs), and sulfur hexafluoride (SF_6), commonly employed in refrigerants, foaming agents, fire extinguishers, solvents, pesticides, and aerosol propellants, possess extended atmospheric lifetimes and impact the climate for several decades (Stocker et al. 2014).

Alongside the anthropogenic activities that are driving climate change, there are a number of major natural factors that influence the climate system, notably variability in solar intensity, volcanic activity, circulation of ocean currents, orbital changes (Milankovitch cycles), melting glaciers and rising sea levels, in addition to natural greenhouse gas emissions (Stocker et al. 2014).

Future Climate Scenarios

The climate modeling community has offered various time-dependent scenarios for climate change, forming the foundation for the majority of future projections in IPCC assessment reports. These standardized scenario sets have undergone evolution with each new generation, as the initial SA90 scenarios were replaced by the IS92 emissions scenarios of the 1990s, which were then succeeded by the Special Report on Emissions Scenarios (SRES) in 2000 and the Representative Concentration Pathways (RCP) in 2010 (Leggett et al. 1992; Nakicenovic et al. 2000; Moss et al. 2010).

The first scenarios namely; SA90, IS92 and SRES are all emission-based scenarios. They start with a set of assumptions based initially on demographic projections. In SRES, they are much more refined, providing a consistent pattern of demographics, international trade, information and technology flows, and other social, technological and economic characteristics of future worlds. Multiple integrated assessment models (IAMs) were used for the SRES, resulting in multiple emissions patterns corresponding to each scenario. However, one scenario for each scheme was chosen as representative "marker" scenario to be used as input to the global models in order to derive the resulting atmospheric concentrations, radiative forcing, and climate change. Thus, the following scenarios could be distinguished: A1FI (fossil fuel intensive), A2 (medium high), B2 (medium low), and B1 (low).

In contrast, RCPs, the most recent set of time-dependent scenarios, is the result of more than two decades of scenario development, and differ from previous standard scenario sets in at least four key respects. Firstly, RCPs are radiative forcing scenarios, and not emission scenarios. Radiative forcing quantifies the additional heat retained by the lower atmosphere due to the increased presence of greenhouse gases, measured in watts per square meter (W/m^2) (Jubb et al. 2013). Each RCP scenario is based on four distinct assumptions about the amount of greenhouse gases expected between 2000 and 2100, giving a likely climate variant related to the chosen emission levels considered as a working hypothesis. The four scenarios are assigned a number according to the change in radiative forcing by the year 2100 compared with pre-industrial levels: RCP 2.6 corresponds to a forcing of + 2.6 W/m^2, RCP 4.5 to + 4.5 W/m^2, and similarly for RCP 6 and RCP 8.5. A higher numerical value indicates a greater amount of energy gained by the Earth-atmosphere system, resulting in more significant warming (van Vuuren et al. 2011; Thomson et al. 2011; Masui et al. 2011; Riahi et al. 2011). The second difference lies in n the fact that, from these radiative forcing values, MAIs are applied in a reverse process to generate a spectrum of emission trajectories and their corresponding technology policies and strategies for each RCP that would have the same final impact on radiative forcing (Cubasch et al. 2013). Thirdly, the PCRs stand out from previous scenarios by incorporating explicit climate policies (for PCR 2.6, PCR 4.5 and PCR 6.0) to limit climate forcing. RCP 8.5 is a scenario in which CO_2 and CH_4 emissions persistently increase, despite a significant reduction in emission growth rates in the second half of the century. Atmospheric CO_2 levels in RCP 8.5 are projected to reach 936 ppm by the end of the century, surpassing 1200 ppm by 2100. Global temperature is projected to rise by 3–5.5 °C in 2100 compared to the 1986–2005 average. In contrast, the RCP 4.5 and

RCP 2.6 scenarios keep atmospheric CO_2 levels below 550 and 450 ppm by 2100 respectively (Collins et al. 2013; Melillo et al. 2014). Fourthly, a variety of socio-economic scenarios were developed independently of the RCPs. A subset of these scenarios has been constrained using emission limitation policies consistent with their underlying storylines, thus forming Shared Socio-Economic Pathways (SSPs) with climate forcing corresponding to RCP values. This combination of SSPs and RCPs is designed to meet the needs of communities responsible for impacts, adaptation and vulnerability (O'Neill et al. 2014).

As consequences, the environmental abiotic and biotic systems will increasingly degrade and the socio-economic component will undergo these transformations, which may cause the scarcity of resources, the upheaval of ecosystems or even crises and problems of insecurity.

Water Vulnerabilities in the Face of Climate Change: Issues and Impacts

Water is a primary natural resource crucial to sustaining life. It's evident that water has a vital abiotic role in maintaining the health of ecosystems and ensuring human survival on this planet, because of its importance to people's lives, agriculture and production processes (Bayart et al. 2010; Layani et al. 2021). Consequently, its crucial socio-economic influence profoundly shapes the well-being of human communities.

Many published investigations have revealed that all sectors of water, whether direct or otherwise, are experiencing the effects of climate change, in particular agriculture, hydropower, the economy, industry, tourism, biodiversity, food security, forestry, health, etc. This confirms the importance of considering water as a crucial aspect of adapting to climate change (Hall et al. 2015; Watts et al. 2015; Alhassan 2021; Wernberg et al. 2016; Zhang et al. 2017; Manes et al. 2021; Abbass et al. 2022).

Given that the repercussions of climate change are mainly water-related phenomena, notably changes in the severity and occurrence of extreme weather events, water will be a key factor in adapting to climate change. In fact, such impacts can be classified into three main categories: water insufficiency (droughts and water shortages), water excess (floods) and contaminated water (pollution) (Wehrey et al. 2023).

In terms of water-related disasters, floods are ranked the most frequent, accounting for 47% of all weather catastrophes, affecting 2.3 billion people worldwide. Meanwhile, storms, although occurring less frequently than floods, were responsible for over 242,000 deaths in the last 21 years, thus being rated the most fatal of all sever weather-related events. Furthermore, even though low-income countries experience only 26% of these natural disasters, an estimated 89% of storm-related fatalities occur within this group of nations (UNISDR 2015; Bhatt et al. 2020).

In addition to these meteorological disasters, there are waves of extreme temperatures, mainly heatwaves, which have been particularly deadly (with an average of 405 deaths per disaster). High-income countries rank second, behind lower-middle-income countries, in terms of the average number of victims per disaster (UNISDR 2015).

On the other hand, climate change-induced warming results in more intense, widespread and persistent extreme weather events, such as droughts (Meehl et al. 2007; Trenberth 2011; Mukherjee et al. 2018).

Climate change is expected to further exacerbate the pressures on available water resources posed by population growth, economic drivers and land-use changes. In fact, water reserves, particularly run-off, are projected to decrease by 10–30% in certain dry regions as a result of lower precipitation and higher evapotranspiration rates. It is also highly plausible that climate change will significantly reduce water resources in all semi-arid zones. Such variations in water availability, driven by climate change, would certainly have adverse consequences for many sectors, including agriculture, water supply, energy production and health (Raneesh 2014).

Water scarcity is complicated by the deterioration of water quality. Indeed, 2 million tons of wastewater as well as industrial and agricultural waste are discharged into waterways. A high portion of this wastewater is released untreated into surface waters (particularly in developing countries), resulting in more than 2 million deaths annually from diarrheal diseases. Additionally, the World Health Organization (WHO) reports that over 80% of all diseases are waterborne (WHO 2014).

Taken together, these data emphasize the key role of water as a multi-purpose resource for human security, environmental sustainability and sustainable development.

Climate Change in Morocco

Geographic Location and Climate Characteristics

Morocco lies in the north-western part of the African continent, surrounded by the Atlantic Ocean to the west and the Mediterranean Sea to the north. Morocco's climate varies widely among regions, as it is influenced by the Atlantic Ocean and plains to the west, the Mediterranean Sea and the Rif Mountains to the north, the Atlas Mountains in the central region, plateaus to the east, and the Sahara Desert to the south and southeast. Consequently, Morocco's climate is heavily shaped by its topography. In fact, most of the country, especially along the coast, is subject to the typical Mediterranean climate, marked by mild, wet winters and hot, dry summers. The northern Mediterranean coastal regions and the southern inland areas are geographically separated in the center by the Atlas Mountains, constituting a natural boundary (USAID 2016; Clement et al. 2021; World Bank Group 2021).

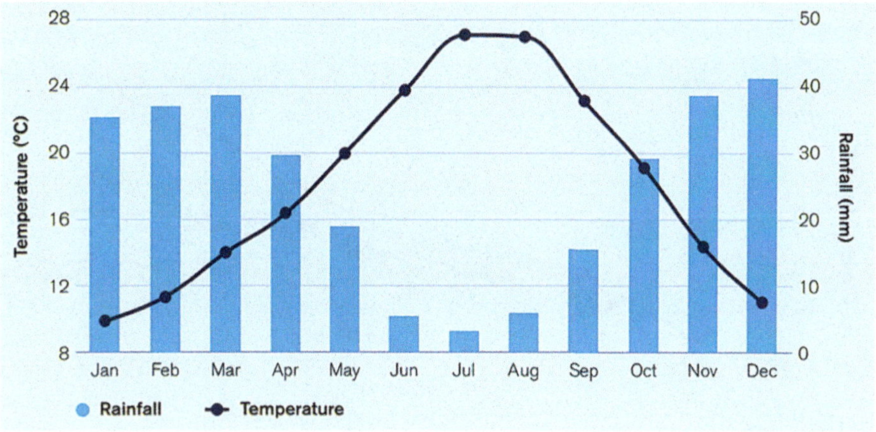

Fig. 1.1 Average monthly temperature and rainfall for Morocco, 1991–2020 (World Bank Group 2021)

The average annual temperature in Morocco is around 17.5 °C, with monthly fluctuations ranging from 9.4 (December, January) to 26 °C (July, August) (Fig. 1.1). However, there is considerable seasonal and regional variability. Temperatures along the coast vary from 18 to 28 °C in summer, and from 8 to 17 °C in winter. Inland, they can reach up to 35 °C in summer and drop below 0 °C in inland mountain areas. The southern zone is much more arid, recording an average annual precipitation of 100 mm. Whereas nationwide, the average annual precipitation is 318.8 mm, with the rainy season extending from November to March, and extremely low precipitation between June and August (Fig. 1.1). The influence of extratropical weather conditions originating from Europe and the Atlantic Ocean introduces cooler air, resulting in a reduction in the rainfall gradient from the northern to the southern part of the country, a phenomenon also influenced by the presence of the Atlas Mountains (Verner et al. 2018; Clement et al. 2021).

Climate Forecasts and Trends in Morocco

Morocco is recognized as one of the main hotspots of climate change due to the acceleration of global warming, notably because of the increase in average temperatures and decrease in precipitation levels in recent decades. Additionally, the country experiences noteworthy inter-annual seasonal variations and shifts in precipitation patterns.

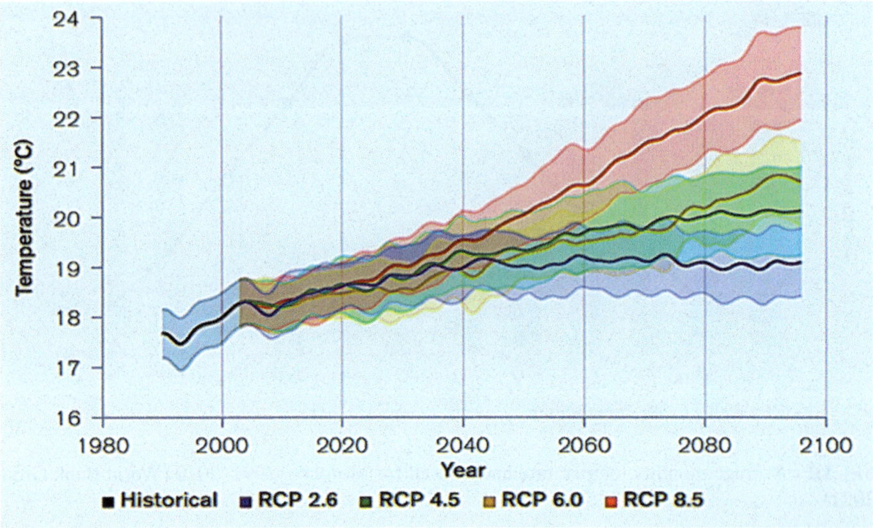

Fig. 1.2 Projected average temperature for Morocco (reference period, 1986–2005) (World Bank Group 2021)

Future Temperature Projections

Temperatures are expected to continue to rise throughout the North African region according to all emission scenarios. Increase of the average annual temperatures is projected to range within 1.5 and 3.5 °C of the mid-century horizon, and to exceed 5 °C by the end of the century (Fig. 1.2). Inland areas are expected to experience heightened warming, particularly with an anticipated increase in the frequency of "hot days" and "hot nights" (Fig. 1.3). The most substantial rise is projected to occur during the months of July, August, and September. Concurrently, there is a forecasted decrease in the occurrence of "cold days" and "cold nights," with their prevalence expected to dwindle to approximately 4% of days and nights by the end of the century (McSweeney et al. 2012).

Rising temperatures are also projected to reduce snow cover in the Atlas Mountains, leading to a reduction in the country's water supply and storage. Heat augmentation and extreme heat conditions will significantly affect human and animal health, agriculture, ecosystems as well as energy production.

Future Precipitation Projections

Precipitation projections for Morocco are widely divergent; however, scenarios forecast a substantial decrease in average annual precipitation throughout the entire country, varying from 10–20 to a 30% reduction in the Saharan region. According to the predictions, the reduction in water resources in Morocco will be due to an

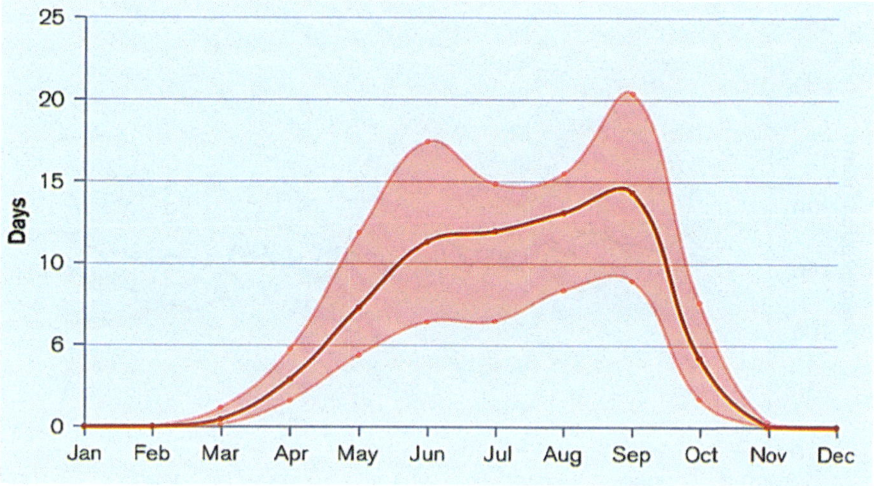

Fig. 1.3 Projected change in hot days (TMax > 35 °C) (RCP 8.5, reference period, 1986–2005) (World Bank Group 2021)

increase in arid periods and droughts (Fig. 1.4). Furthermore, even if precipitation remains unchanged, the elevated temperatures will lead to an increase in evaporation, causing a decline in surface soil moisture. Thus, the available irrigation water sourced from rivers and mountain streams will face escalating challenges, contributing to pronounced water shortages in the arid southern regions (McSweeney et al. 2012; World Bank Group 2021).

Perceived and Potential Effects of Climate Change in Morocco

Water Availability and Supply

Morocco's water resources are unevenly distributed throughout the country. In fact, the coastal plains are frequently flooded, while the south experiences water shortages throughout the year. The availability of renewable water resources per capita has fallen by around 60% since 1960, as a result of non-climate-related stress drivers including population growth in the north, irrigation expansion, and urban, industrial and tourism development. Meanwhile, rising temperatures and irregular rainfall have reduced river flows and accelerated evaporation and silting of storage dams, causing a 20% reduction in global water resources over the latest 30 years (Verner et al. 2018). These alterations add complexity to the management of dams and flood control. Thus, maintaining a steady supply of water to ensure a sustainable socio-economic development of the country is becoming a long-term concern (WWC 2018).

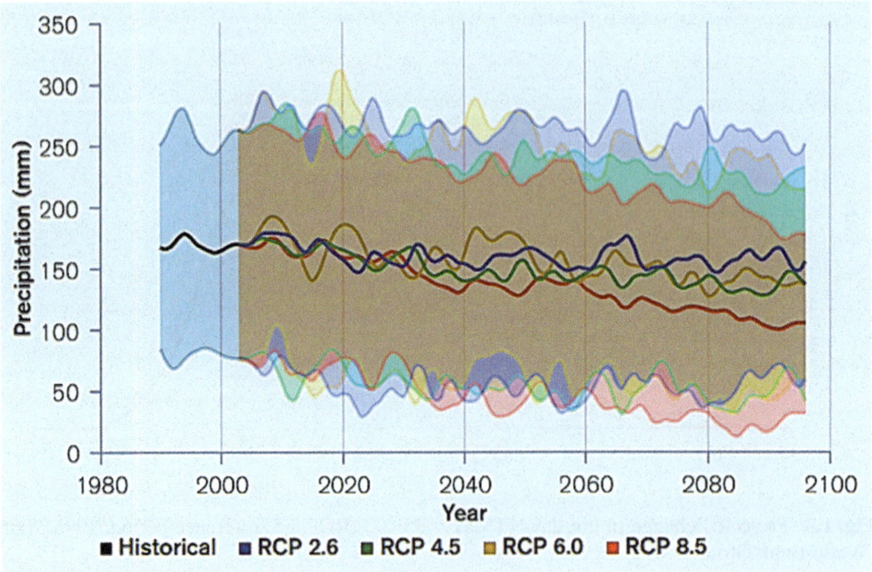

Fig. 1.4 Projected annual average precipitation in Morocco (reference period, 1986–2005) (World Bank Group 2021)

Drought is a natural recurring phenomenon, and has been experienced over the years. Nevertheless, current climate change forecasts predict an increase in the frequency, intensity and duration of droughts, with significant repercussions in the food, water, energy and health sectors (Godde et al. 2021). Morocco is also subject to more pronounced hydro-meteorological extremes, with locally more intense precipitation and a higher frequency of intense flooding causing significant economic losses and property damage. With climate change, demand for irrigation will increase, consuming around 80–85% of available water, although this will differ depending on the area and the allocation of irrigation water. Climate change will significantly alter fertile zones, transforming them from semi-arid to arid and from sub-humid to semi-arid. These zones, already in decline since the 1970s, are projected to continue diminishing. The temperature increase has led to water shortages from 2020, especially in the south, by reducing river flow. Additionally, the reduction of mountain snow cover will increase the flooding risk in October and November, affecting seasonal water availability. Irregular precipitation and intensification of drought will accelerate spring thaw, leading to a reduction in snow reserves in plains, an increase in irrigation demand, and promoting dam siltation due to heavy rainfall and riverbank erosion (USAID 2016; World Bank Group 2021).

The water storage capacity of all existing dams has exhibited declines, reaching 20% against the period 1970–2005 and a more substantial 35% decrease when compared with the 1940–1970 timeframe (Aoubouazza et al. 2013). Projections indicate that the Hassan Addahkhil and Idriss I reservoirs, crucial water sources, are anticipated to experience reductions ranging from 7 to 40% by the 2080s (Verner et al. 2018). Presently, 18% of cultivated lands rely on irrigation. Escalating water stress is bound to result in the excessive exploitation of groundwater resources, a phenomenon already unfolding in several regions. Another major concern, especially for water managers and decision-makers, revolves around the depletion of groundwater reserves. Current estimates indicate an annual extraction of approximately 5 billion m^3 of water, compared with a renewable potential of 4 billion m^3 (Hssaisoune et al. 2020). Similarly, Aoubouazza et al. (2013) reported an alarming drop in piezometric levels in most aquifers; a 40-m decline in the Souss region, a 30-m drop in the Haouz region, and a substantial 60-m decrease in the Saïss region. This swift decline in groundwater levels, averaging 1.5 billion m^3 annually, is commonly linked to insufficient groundwater recharge and excessive expansion of agricultural activities (Bouchaou et al. 2011; Ait Kadi and Ziyad 2018). This excessive utilization has lowered piezometric levels, diminished spring flows and degraded groundwater quality due to pollution, whose main sources are agriculture, industry and households (DRPE 2014; Verner et al. 2018). Moreover, Morocco's Mediterranean coast and the Moulouya river delta, both of which are of great economic and ecological importance, are endangered by rising sea levels and the associated coastal erosion and saline intrusion (Snoussi et al. 2008; World Bank Group 2021). The deterioration in water quality is worsened by fluctuations in hydrological regimes, in particular the reduction in water flows during the dry summer season.

Agriculture and Food Security

Agriculture, the primary source of income for impoverished rural communities, is the most sensitive socio-economic sector to climate fluctuations (Ortiz-Bobea 2021). Research confirms that climate change will have adverse implications for agriculture, thereby placing global food security under serious risk, especially for vulnerable populations in emerging nations (Ziervogel and Zermoglio 2009; Stocker et al. 2014; Nelson et al. 2009; USEPA 2017). Certainly, the current and expected fall in annual precipitation coupled with escalating temperatures and heightened spatiotemporal irregularities, will intensify water scarcity in numerous warmer regions. This, in turn, is bound to have a profound impact on agricultural production (Abdelmajid et al. 2021).

In Morocco, agriculture is also of crucial interest to the economy, food security and rural livelihoods. It is responsible for 13–14% of the national GDP. As regards the national food security, agriculture covers a large portion of Moroccans' food needs, primarily vegetables, fruits, dairy products and meat. In social terms, agriculture employs nearly 40% of the working population nationwide, and 74% in rural areas (Ministry of Economy 2019). Nevertheless, this vital socio-economic sector is facing challenges posed by demographic pressure, erratic rainfall and land degradation (MAFRDWF 2020). Most agricultural production (87%) depends on rainfall and as a result is vulnerable to increasing variability (Achli et al. 2022). In recent decades, the consequences of climate change on Morocco's natural resources have been felt (Ouraich 2015). Investigations on the impact of climate change on agriculture indicate that crops will be especially affected by drought. Cereal production, for instance, varies from 100 million quintals per year in good rainfall periods to 18 million quintals annually in dry times with respective yields fluctuating between 17 and 4 quintals per ha. In fact, dry periods seem to exert an overriding influence on the growth phase (Aoubouazza et al. 2013; World Bank Group 2021). Extreme weather conditions, notably drought, have already caused significant declines in yields, especially a 70% drop in winter cereal yields for 2016 compared to 2015 (Verner et al. 2018). Projections indicate an increase in crop water requirements (up to 12%), putting further pressure on limited water resources. The forecast impacts involve yield reductions of up to 75% due to higher temperatures and lower rainfall in dry years. Reduced water availability also affects irrigation, compromising the viability of irrigated agriculture. Despite a high theoretical allocation of water, the actual average allocation in recent years was 58% of this amount, highlighting difficulties in water management in Morocco (USAID 2016). Available data also suggests that climate change (drought conditions) could lead to disappearance of certain crops and trees, emergence of new diseases, and disruption of agriculture-related activities (i.e., livestock breeding and production) (Aoubouazza et al. 2013; World Bank Group 2021).

To address this alarming situation, Morocco needs to step up its efforts and put in place a comprehensive and appropriate policy and risk management framework to mitigate the risks of climate change. This includes establishing mechanisms for strategic monitoring and financing of adaptation and mitigation measures.

Conclusions

The major challenges posed by climate change increase the urgency of adaptation and vulnerability reduction measures. In Morocco, policies concerning both sectors—water and agriculture—were initiated many years ago, and are currently being pursued through the "Water Highway Project". However, there is a pressing need to further strengthen national policies and plans aimed to increase the country's resilience and adaptability to climate events and disasters. Such improvement requires measures directed towards preserving productive assets and exploiting the range of technological innovations. Furthermore, it is essential to set up schemes that meet both the specific needs of the agricultural sector and socio-economic developments, in order to ensure social equality and stability.

References

Abbass K, Qasim MZ, Song H, Murshed M, Mahmood H, Younis I (2022) A review of the global climate change impacts, adaptation, and sustainable mitigation measures. Environ Sci Pollut Res 29(28):42539–42559

Abdelmajid S, Mukhtar A, Baig MB, Reed MR (2021) Climate change, agricultural policy and food security in Morocco. In: Emerging challenges to food production and security in Asia, Middle East, and Africa: climate risks and resource scarcity, pp 171–196

Achli S, Epule TE, Dhiba D, Chehbouni A, Er-Raki S (2022) Vulnerability of barley, maize, and wheat yields to variations in growing season precipitation in Morocco. Appl Sci 12(7):3407

Agrawala S (1998) Context and early origins of the Intergovernmental Panel on Climate Change. Clim Change 39(4):605–620

Ait Kadi M, Ziyad A (2018) Integrated water resources management in Morocco. In: Global water security: lessons learnt and long-term implications, pp 143–163

Alhassan H (2021) The effect of agricultural total factor productivity on environmental degradation in sub-Saharan Africa. Sci Afr 12:e00740

Aoubouazza M, Rajel R, Essafi R (2013) Impact of extreme climate events on water resources and agriculture and biodiversity in Morocco. J Climatol Weather Forecast 1(104):2

Bahir M, Ouhamdouch S, Ouazar D, El Moçayd N (2020) Climate change effect on groundwater characteristics within semi-arid zones from western Morocco. Groundw Sustain Dev 11:100380

Bayart JB, Bulle C, Deschênes L, Margni M, Pfister S, Vince F, Koehler A (2010) A framework for assessing off-stream freshwater use in LCA. Int J Life Cycle Assess 15:439–453

Bazzaz FA, Sombroek WG (eds) (1996) Global climate change and agricultural production: direct and indirect effects of changing hydrological, pedological, and plant physiological processes. Food & Agriculture Organization

Bennouna A (2020) Gestion de l'eau au Maroc et changement climatique. Espace Géogr Soc Maroc (32)

Bhatt M, Patel RB, Gleason K, Pandya M (2020) Disaster risk reduction and climate change adaptation in South Asia. In: Oxford research encyclopedia of natural hazard science. https://doi.org/10.1093/acrefore/9780199389407.013.345

Bouaicha R, Benabdelfadel A (2010) Variabilité et gestion des eaux de surface au Maroc. Sécheresse 21(1e):1–5

Bouchaou L, Tagma T, Boutaleb S, Hssaisoune M, El Morjani ZEA (2011) Climate change and its impacts on groundwater resources in Morocco: the case of the Souss-Massa basin. In: Climate change effects on groundwater resources: a global synthesis of findings and recommendations, p 129

Clement V, Rigaud KK, de Sherbinin A, Jones B, Adamo S, Schewe J, Sadiq N, Shabahat E (2021) Groundswell. Acting on internal climate migration. Part II. The World Bank, Washington, DC. https://doi.org/10.1596.36248

Collins M, Knutti R, Arblaster J, Dufresne JL, Fichefet T, Friedlingstein P et al (2013) Long-term climate change: projections, commitments and irreversibility

Cubasch U, Wuebbles D, Chen D, Facchini MC, Frame D, Mahowald N, Winther JG (2013) Introduction. In: Stocker TF, Qin D, Plattner GK, Tignor M, Allen SK, Boschung J, Nauels A, Xia Y, Bex V, Midgley PM (eds) Climate change 2013: the physical science basis. Contribution of working group I to the fifth assessment report of the Intergovernmental Panel on Climate Change. Cambridge University Press, Cambridge, New York, NY, 119–158. https://doi.org/10.1017/CBO9781107415324.007

DRPE [Direction de la Recherche et de la Planification de l'Eau] (2014) Eaux Souterraines Au Maroc, Comment Concilier Satisfaction Des Besoins et Développement Durable Des Ressources En Eau Souterraines, unpublished report. Ministère délégué auprès du Ministre de l'Energie, des Mines, de l'Eau et de l'Environnement Chargé de l'Eau, Rabat, pp 1–53

El Ajhar L, El Khachine D, El Bakouri A, El Kharrim K, Belghiti D (2018) Evolution of rainfall from 1960 to 2015 in Morocco. Int J Res Sci Manage 5(10):47–56

Ellison D (2018) Forests and water. In: Background analytical study, p 2
Fakana ST (2020) Causes of climate change. Glob J Sci Front Res 20:7–12
FAO [Food and Agriculture Organization] (2011) "Energy-smart" food for people and climate: issue paper. Food and Agriculture Organization of the United Nations. http://www.fao.org/3/i2454e/i2454e00.pdf
Godde CM, Mason-D'Croz D, Mayberry DE, Thornton PK, Herrero M (2021) Impacts of climate change on the livestock food supply chain; a review of the evidence. Glob Food Secur 28:100488
Güçlü Y (2014) Sea surface temperature anomalies at the Mediterranean coast of Turkey (period of 1968–2010). Procedia Soc Behav Sci 120:476–486
Haghipour N, Burg JP (2014) Geomorphological analysis of the drainage system on the growing Makran accretionary wedge. Geomorphology 209:111–132
Hall CM, Amelung B, Cohen S, Eijgelaar E, Gössling S, Higham J, Leemans R, Peeters P, Ram Y, Scott D (2015) On climate change skepticism and denial in tourism. J Sustain Tour. https://doi.org/10.1080/09669582.2014.953544
Hirich A, Fatnassi H, Ragab R, Choukr-Allah R (2016) Prediction of climate change impact on corn grown in the south of Morocco using the Saltmed model. Irrig Drain 65(1):9–18
Hssaisoune M, Bouchaou L, Sifeddine A, Bouimetarhan I, Cheh-bouni A (2020) Moroccan groundwater resources and evolution with global climate changes. Geosciences 10(2):81
IPCC [Intergovernmental Panel on Climate Change] (2019) Climate change and land: an IPCC special report on climate change, desertification, land degradation, sustainable land management, food security, and greenhouse gas fluxes in terrestrial ecosystems. In: Shukla PR, Skea J, Calvo Buendia E, Masson-Delmotte V, Pörtner H-O, Roberts DC, Zhai P, Slade R, Connors S, van Diemen R, Ferrat M, Haughey E, Luz S, Neogi S, Pathak M, Petzold J, Portugal Pereira J, Vyas P, Huntley E, Kissick K, Belkacemi M, Malley J (eds). In press
Jubb I, Canadell P, Dix M (2013) Representative concentration pathways (RCPs). Australian Government, Department of the Environment
Kumar P, Tokas J, Kumar N, Lal M, Singal HR (2018) Climate change consequences and its impact on agriculture and food security. Int J Chem Stud 6(6):124–133
Layani G, Bakhshoodeh M, Zibaei M, Viaggi D (2021) Sustainable water resources management under population growth and agricultural development in the Kheirabad river basin, Iran. Bio-Based Appl Econ 10(4):305–323
Leggett J, Pepper WJ, Swart RJ, Edmonds J, MeiraFilho LG, Mintzer I, Wang MX (1992) Emissions scenarios for the IPCC: an update. Clim Change 1040:75–95
MAFRDWF [Ministry of Agriculture, Fisheries, Rural Development, Water and Forest] (2020) Le Plan Maroc Vert. Bilan et impacts, 2008–2018. https://www.agriculture.gov.ma/fr/actualites/publication-dun-rapport-sur-le-bilan-et-impacts-du-plan-maroc-vert
Manes S, Costello MJ, Beckett H, Debnath A, Devenish-Nelson E, Grey KA et al (2021) Endemism increases species' climate change risk in areas of global biodiversity importance. Biol Conserv 257:109070
Mason N, Nalamalapu D, Corfee-Morlot J (2019) Climate change is hurting Africa's water sector, but investing in water can pay off
Masui T, Matsumoto K, Hijioka Y, Kinoshita T, Nozawa T, Ishiwatari S et al (2011) An emission pathway for stabilization at 6 Wm^{-2} radiative forcing. Clim Change 109:59–76
McSweeney C, New M, Lizcano G (2012) UNDP climate change country profiles—Morocco. https://www.geog.ox.ac.uk/
Meehl GA, Stocker TF, Collins WD, Friedlingstein P, Gaye AT, Gregory JM et al (2007) Climate change 2007: the physical science basis. Contribution of working group I to the fourth assessment report of the Intergovernmental Panel on Climate Change, vol 32. Cambridge University Press, pp 10013–12473
Melillo JM, Richmond TT, Yohe G (2014) Climate change impacts in the United States. In: Third national climate assessment, U.S. global change research program, vol 52, pp 150–174

Ministry of Economy (Morocco) (2019) The Moroccan agricultural sector: structural trends, challenges and development prospects (in French). DEPF. https://www.finances.gov.ma/Publication/depf/2019/Le%20secteur%20agricole%20marocain.pdf

Moss RH, Edmonds JA, Hibbard KA, Manning MR, Rose SK, Van Vuuren DP et al (2010) The next generation of scenarios for climate change research and assessment. Nature 463(7282):747–756

Mukherjee S, Mishra A, Trenberth KE (2018) Climate change and drought: a perspective on drought indices. Curr Clim Change Rep 4:145–163

Nakicenovic N, Alcamo J, Davis G, de Vries B, Fenhann J, Gaffin S et al (2000) IPCC special report on emissions scenarios. Cambridge, New York, NY

Nelson GC, Rosegrant MW, Koo J, Robertson R, Sulser T, Zhu T, Ringler C, Msangi S, Palazzo A, Batka M, Magalhaes M, Valmonte-Santos R, Ewing M, Lee D (2009) Climate change: impact on agriculture and costs of adaptation, vol 21. International Food Policy Research Institute

O'Neill BC, Kriegler E, Riahi K, Ebi KL, Hallegatte S, Carter TR, Mathur R, Van Vuuren DP (2014) A new scenario framework for climate change research: the concept of shared socioeconomic pathways. Clim Change 122:387–400

Omer AM (2008) Energy, environment and sustainable development. Renew Sustain Energy Rev 12(9):2265–2300

Oreggioni GD, Ferraio FM, Crippa M, Muntean M, Schaaf E, Guizzardi D et al (2021) Climate change in a changing world: socio-economic and technological transitions, regulatory frameworks and trends on global greenhouse gas emissions from EDGAR v. 5.0. Glob Environ Change 70:102350

Ortiz-Bobea A (2021) The empirical analysis of climate change impacts and adaptation in agriculture. In: Handbook of agricultural economics, vol 5. Elsevier, pp 3981–4073

Ouhamdouch S, Bahir M, Carreira PM (2018) Impact du changement climatique sur la ressource en eau en milieu semi-aride: exemple du bassin d'Essaouira (Maroc). Rev Sci Eau 31(1):13–27

Ouraich I (2015) Agriculture, climate change, and adaptation in Morocco: a computable general equilibrium analysis. Doctoral dissertation, Purdue University

Pielke RA Jr (2004) What is climate change? Energy Environ 15(3):515–520

Raneesh KY (2014) Impact of climate change on water resources. J Earth Sci Clim Change 5(3):1

Riahi K, Rao S, Krey V, Cho C, Chirkov V, Fischer G et al (2011) RCP 8.5—a scenario of comparatively high greenhouse gas emissions. Clim Change 109:33–57

Rochdane S, Bounoua L, Zhang P, Imhoff ML, Messouli M, Yacoubi-Khebiza M (2014) Combining satellite data and models to assess vulnerability to climate change and its impact on food security in Morocco. Sustainability 6(4):1729–1746

Schilling J, Hertig E, Tramblay Y, Scheffran J (2020) Climate change vulnerability, water resources and social implications in North Africa. Reg Environ Change 20:1–12

Sharma K, Mehta S, Sinha AK (eds) (2012) Global warming, human factors and environment: anthropological perspectives. Excel India Publishers

Simonneaux V, Cheggour A, Deschamps C, Mouillot F, Cerdan O, Le Bissonnais Y (2015) Land use and climate change effects on soil erosion in a semi-arid mountainous watershed (High Atlas, Morocco). J Arid Environ 122:64–75

Sims REH, Goresvski V, Anenberg S (2015) Black carbon mitigation and the role of the global environment facility: a STAP advisory document. GEF, Washington, DC. http://www.stapgef.org/node/1607

Snoussi M, Ouchani T, Niazi S (2008) Vulnerability assessment of the impact of sea-level rise and flooding on the Moroccan coast: the case of the Mediterranean eastern zone. Estuar Coast Shelf Sci 77(2):206–213

Stocker TF, Qin D, Plattner GK, Tignor MM, Allen SK, Boschung J et al (2014) Climate change 2013: the physical science basis. In: Contribution of working group I to the fifth assessment report of the Intergovernmental Panel on Climate Change

Strasser U, Vilsmaier U, Prettenhaler F, Marke T, Steiger R, Damm A et al (2014) Coupled component modelling for inter- and transdisciplinary climate change impact research: dimensions of integration and examples of interface design. Environ Model Softw 60:180–187

Tahir AA, Chevallier P, Arnaud Y, Neppel L, Ahmad B (2011) Modeling snowmelt-runoff under climate scenarios in the Hunza River basin, Karakoram Range, Northern Pakistan. J Hydrol 409(1–2):104–117

Thomson AM, Calvin KV, Smith SJ, Kyle GP, Volke A, Patel P et al (2011) RCP 4.5: a pathway for stabilization of radiative forcing by 2100. Clim Change 109:77–94

Trabelsi R, Zouari K (2019) Coupled geochemical modeling and multivariate statistical analysis approach for the assessment of groundwater quality in irrigated areas: a study from North Eastern of Tunisia. Groundw Sustain Dev 8:413–427

Tramblay Y, Ruelland D, Bouaicha R, Servat E (2014a) Projected climate change impacts on water resources in northern Morocco with an ensemble of regional climate models. Proc Int Assoc Hydrol Sci 363:250–255

Tramblay Y, Somot S, Zsolt Torma C, Coppola E, Romera R, Domínguez M, Ángel Gaertner M, Ruelland D, Bouaicha R, Servat E (2014b) Evaluation of Med-CORDEX regional climate model simulations for hydrological impact studies in Morocco. In: EGU general assembly conference abstracts, p 5713

Trenberth KE (2011) Changes in precipitation with climate change. Clim Res 47(1–2):123–138

UNISDR [United Nations Office for Disaster Risk Reduction] (2015) The human cost of weather-related disasters 1995–2015. UNISDR, Centre for Research on the Epidemiology of Disasters CRED, Geneva, p 30

USAID [Unites States Agency for International Development] (2016) Climate change risk profile—Morocco. https://www.climatelinks.org/sites/default/files/asset/document/2016_USAID_Climate%20Risk%20Profile%20-%20Morocco.pdf

USEPA (2017) Climate impacts on agriculture and food supply. United States Environmental Protection Agency, Washington, DC

Van Vuuren DP, Stehfest E, den Elzen MG, Kram T, van Vliet J, Deetman S et al (2011) RCP 2.6: exploring the possibility to keep global mean temperature increase below 2°C. Clim Change 109:95–116

Verner D, Treguer D, Redwood J, Christensen J, McDonnell R, Elbert C, Konishi Y, Belghazi S (2018) Climate variability, drought, and drought management in Morocco's agricultural sector. World Bank

Watts N, Adger WN, Agnolucci P, Blackstock J, Byass P, Cai W et al (2015) Health and climate change: policy responses to protect public health. Lancet 386(10006):1861–1914

Weart SR (2008) The discovery of global warming: revised and, expanded. Harvard University Press

Wehrey F, Dargin J, Mehdi Z, Muasher M, Yahya M, Kayssi I et al (2023) Climate change and vulnerability in the Middle East, Nov 2023. https://policycommons.net/artifacts/4430668/climate-change-and-vulnerability-in-the-middle-east/5227290/. CID: 20.500.12592/7vg9ts

Wernberg T, Bennett S, Babcock RC, De Bettignies T, Cure K, Depczynski M et al (2016) Climate-driven regime shift of a temperate marine ecosystem. Science 353(6295):169–172

WHO (2014) Water. World Health Organization. http://www.who.int/topics/water/en/

Woillez MN (2019) Revue de littérature sur le changement climatique au Maroc: observations, projections et impacts. Papiers de recherché, pp 1–33

World Bank Group (2021) Climate risk country profile: Morocco. https://climateknowledgeportal.worldbank.org/sites/default/files/2021-09/15725-WB_Morocco%20Country%20Profile-WEB.pdf

WWC [World Water Council] (2018) Global water security: lessons learnt and long-term implications. Springer

Zhang M, Liu N, Harper R, Li Q, Liu K, Wei X et al (2017) A global review on hydrological responses to forest change across multiple spatial scales: importance of scale, climate, forest type and hydrological regime. J Hydrol 546:44–59

Ziervogel G, Zermoglio F (2009) Climate change scenarios and the development of adaptation strategies in Africa: challenges and opportunities. Clim Res 40(2–3):133–146

Chapter 2
Are the Heat Waves Recorded in Morocco During the Summer of 2022 Exceptional?

Jaafar El Kassioui and Mohamed Hanchane

Abstract This study deals with the heat waves recorded in June and July 2022 in Morocco, in comparison with the chosen reference period 1980–2019. The objective of this study is to clarify whether the heat waves of the year are exceptional in terms of record values, duration and frequency of heat waves in our study area. The study is based on the daily data of the maximum temperature (Tmax) observed. The identification of heat waves is based on the criterion that the Tmax exceed the 90th percentile for at least 3 consecutive days. Our results show that the extreme values associated with the heat waves of the months of June and July of the year 2022 did not exceed the records of the reference period 1980–2019 for all the stations analyzed. On the other hand, records for the duration of heat waves exceeded those of the reference period 1980–2019. In addition, all the stations studied recorded a large annual number of heat waves and longer durations during the second half of the reference period.

Keywords Heat wave · Summer · Absolute record · Morocco · June and July 2022

Introduction

The Intergovernmental Panel on Climate Change (IPCC) classified heat waves as extreme weather events. This phenomenon is expected to increase in the future in terms of frequency and severity (IPCC 2019). McGregor et al. (2015) defined a heat wave as "periods of unusually hot dry or hot humid weather that have an insidious onset and cessation, a duration of at least 23 days with a discernible impact on human and natural systems as demonstrated clearly by a number of devastating heat wave events over the last decade". Studies carried out in different regions of the world show that heat waves and hottest days become more frequent and intense (Founda and Santamouris 2017; Kueh et al. 2017; Wie et al. 2021; Awasthi et al.

J. El Kassioui (✉) · M. Hanchane
Department of Geography, Laboratory Territory Heritage and History, FLSH Dhar El-Mehraz, Sidi Mohamed Ben Abdellah University, 30050 Fez, Morocco
e-mail: Jaafar.elkassioui@usmba.ac.ma

K. Kahime et al. (eds.), *Climate Change Effects and Sustainability Needs*, Springer Climate, https://doi.org/10.1007/978-3-031-59603-2_2

2022; Founda et al. 2022; Serrano-Notivoli et al. 2022; Zhang et al. 2022). In this study, a heat wave is defined when Tmax values exceed the 90th percentile for three or more consecutive days. This criterion remains the most usable in climate studies (Tomczyk and Sulikowska 2018).

Overall, the summer months of the year 2022 were marked by a very hot summer that attracted worldwide attention (Lu et al. 2023). The maximum temperature (Tmax) was more severe in several climatic regions in the world, including Europe, North Africa, North America and Southeast Asia (Lu et al. 2023). The abnormality of the summer 2022 heat waves has led many studies to focus on their analyzes and explanations (Lu et al. 2023; Jiang et al. 2023; Tobías et al. 2023; Yule et al. 2023; Zhang et al. 2023). Indeed, summer 2022 Tmax exceeded records in several European regions (Yule et al. 2023) where at least 15,000 recorded deaths were specifically caused by this heat wave (Henri and Kluge 2022).

In Morocco, the effect of heat has negatively affected forest ecosystems, water resources and agricultural production. The results of Lu et al. (2023) show that Morocco is among the regions that experienced a positive maximum temperature anomaly during the summer of 2022. If particular weather conditions related to the hot winds from east (*Chergui*) are responsible for heat waves, the question is to know the degree of their exceptionality compared to a reference period. To answer this question, we use a statistical approach based on the daily Tmax data for the months of June and July. The objective of the study is to show the location of the heat waves that Morocco experienced during the months of June and July 2022 in the context of climate change. Because any increase of the frequency and duration of the heat waves can cause serious consequences on many living sectors. Including human health and safety, as demonstrated by numerous studies around the world (Parente et al. 2018; Cheng et al. 2019; Stillman 2019; Basarin et al. 2020).

The Study Area

The geographical position of Morocco in the North-West of the African continent (Fig. 2.1) is favorable to the advent of heat waves, especially during the summer season. Indeed, the geographical position of Morocco makes it under the influence of the subtropical Azores anticyclone and a warm air ridge at altitude, especially during the summer. This situation is favorable to the installation of hot meteorological conditions which are amplified under the effect of the Saharan depression. In association with the Azores ridge in the north of the country, the depression directs hot and dry easterly winds over the country which further increase the Tmax.

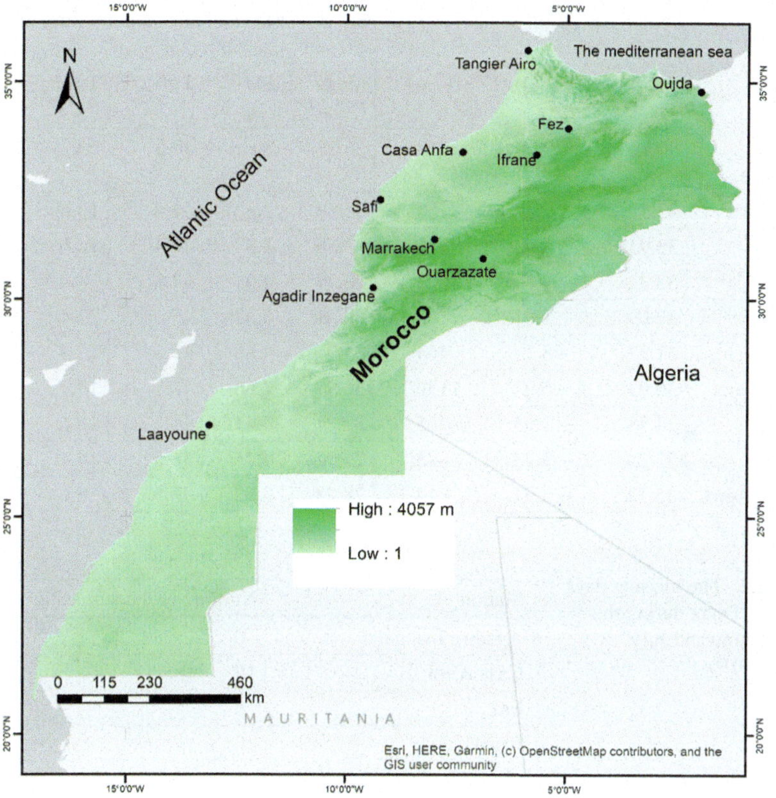

Fig. 2.1 Geographical location of the stations studied

Data

In this study, we used maximum daily temperature (Tmax) data from the National Department of Meteorology (DNM). These data concern the months of June and July of the year 2022 relating to ten stations (Table 2.1). In addition, we use the daily Tmax of a reference period 1980–2019 of the same month (June and July). The total sample for June and July equals 2440 observations. We have some missing data which is corrected by the average of the day before and after the missing data (Table 2.2).

The choice of data is based on the geographical representativeness of three main factors:

- The latitudinal gradient (North/South);
- The differentiation between the coast and the mainland;
- The altitudinal gradation.

Table 2.1 Geographical position and descriptive statistics of the climatic stations studied for daily Tmax

Station	X (°N)	Y (°O)	H (m)	Mean (°C)	SD	Tmax (°C)	Tmin (°C)	90th (°C)
Agadir Inzegane	30.26	− 9.39	18	26.17	3.8	48.5	20.2	29.4
Casa Anfa	33.3	− 7.38	56	25.58	2.5	44.4	19.6	28.1
Fez Sais	33.93	− 4.98	571	33.38	4.87	46.4	16.6	39.8
Ifrane	33.32	− 5.7	1663	28.27	4.6	37.7	8.9	33.4
Laayoune	27.09	− 13.12	63	29.31	3.32	47	22.4	33.2
Marrakech	31.37	− 8	464	35.9	4.66	49.6	17.4	42.3
Ouarzazate	30.93	− 6.9	1136	36.99	2.66	42.8	19.7	39.9
Oujda	34.78	− 1.93	465	32.81	4.41	46.2	16.6	38.3
Safi	32.2	− 9.23	43	27.92	4.46	46.5	14.1	33.9
Tanger Aero	35.72	− 5.91	14	28.28	3.92	43.5	18.6	33.7

Table 2.2 The missing days for daily Tmax during the months June and July (1980–2019)

Station	Number of days missing
Agadire Inzegane	0
Casa Anfa	17
Fez	2
Ifrane	0
Laayoune	0
Marrakech	2
Ouarzazate	0
Oujda	0
Safi	2
Tangier Airo	5

Method

In this study, we defined heat waves when Tmax exceeds the 90th percentile for at least three consecutive days (Perkins 2015; Zhang et al. 2005). The use of the 90th allows us to have a high number of heat waves (Perkins and Alexander 2013).

The availability of daily Tmax for a reference period 1980–2019 made it possible to calculate the climate indices recommended by "the team of experts on the sectoral climate index (ET-SCI: https://climpact-sci.org/indexes/)" using the Climpact application which is developed in R. The indices used are Tx50 and Tx90 which correspond to days where the percentage of Tmax above the 50th median and the 90th percentile successively (Peterson et al. 2001; Nairn and Fawcett 2011). These indices were calculated for the summer season (JJA) and the annual average. The time series of both indices were used to analyze their trends. So, if there is an upward trend for these two indicators, it is likely that heat waves will become more pronounced over time. Indeed, we rely on all the summer months in order to also know the seasonal variation days greater than or equal to Tx50 and Tx90 for the Tmax, especially since the summer season tends to increase more than the season winter (Born et al. 2008), and we compare this season (JJA) with the annual evolution in order to know its weight.

Afterwards, we analyzed the heat waves for the months of June and July 2022 by comparing them to the reference period 1980–2019 for the same months (June and July). This comparison concerned the duration and the overtaking of the record values. For example, the Fez station recorded the maximum duration of the reference heatwave, which lasted 8 days (between July 11 and 18, 2017), while the maximum heatwave duration was recorded in the current period, which lasted 10 days (between July 8 and 17, 2022).

First, we extract all days exceeding the 90th threshold (Table 2.1) from the total number of observations of 2440 days at each station for the reference period of June and July (1980–2019) for the daily maximum temperature. Second, we only retain observations that occupy at least three consecutive days greater than or equal to the 90th. Finally, we extract the heat waves for the months of June and July 2022 from this reference threshold, that is to say they are here applied to 61 observations at each station.

Results and Discussion

Trend of the Two Indices Tx50 and Tx90

In this first step, we studied the trends of two climatic indices Tx50 and Tx90 which are calculated from daily maximum temperature data based on the Climpact computer application. After determining the two indices, the calculation of trends is based on the slope estimator of Theil-Sen (Sen 1968). This nonparametric method of trend estimation is robust to outliers and nonnormality based on Kendall's rank correlation (Winslow et al. 2015).

The results obtained show a trend towards an increase in the two indices Tx50 and Tx90 in all the stations studied (Figs. 2.2 and 2.3). The tendencies towards the increase of the summer season are stronger compared to the annual scale, which means that this season is among the seasons which evolve towards the increase more strongly than the other seasons.

During the summer, the rate of increase in the number of hot days exceeding the 50th per decade records its minimum at the Tangier-Airo station (4.29 days per 10 years) and its maximum at the Ouarzazate station (12.73 days per 10 years). Also, on an annual scale, the rate of increase in the number of hot days that exceeds the 50th per decade recorded its minimum at the Ifrane station (3.2 days per 10 years)

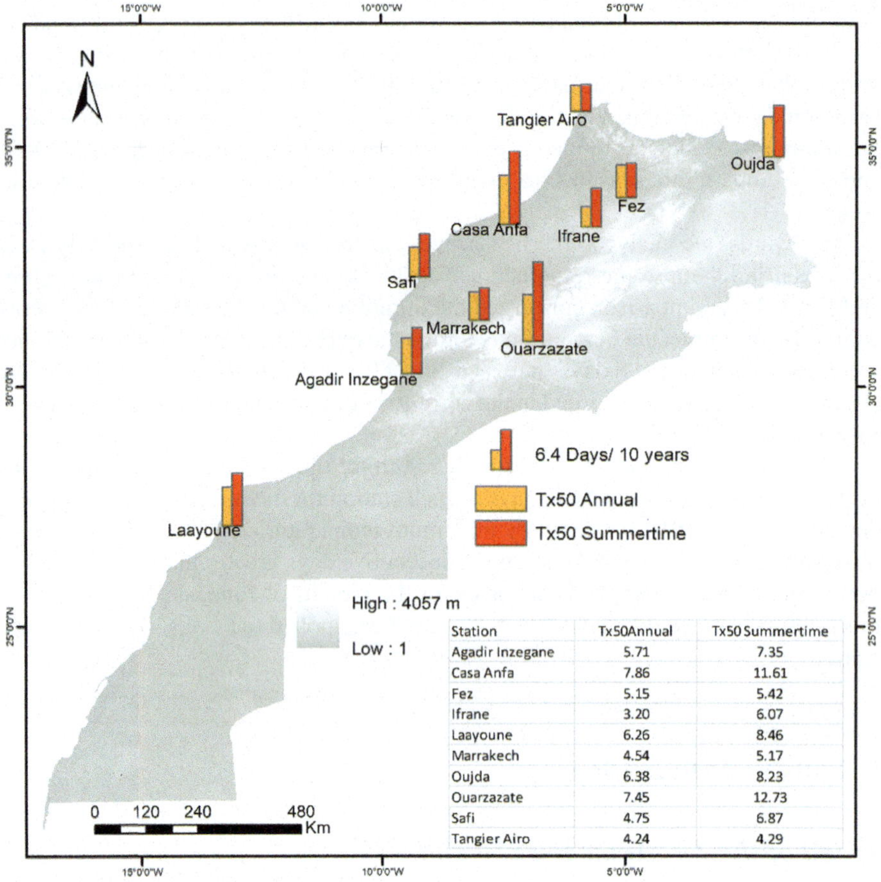

Station	Tx50Annual	Tx50 Summertime
Agadir Inzegane	5.71	7.35
Casa Anfa	7.86	11.61
Fez	5.15	5.42
Ifrane	3.20	6.07
Laayoune	6.26	8.46
Marrakech	4.54	5.17
Oujda	6.38	8.23
Ouarzazate	7.45	12.73
Safi	4.75	6.87
Tangier Airo	4.24	4.29

Fig. 2.2 The rate of increase in hot days per decade according to the Tx50th index (orange color shows the annual scale, red color shows the summer season (JJA))

and its maximum at the Casa-Anfa station (7.86 days per 10 years). As for the rate of increase in the number of hot days exceeding the 90th per decade, it is moving upwards at a slower pace. Indeed, during the summer season, the minimum rate is recorded at the Safi station (1.29 days per 10 years) and the maximum rate is recorded at the Ouarzazate station (6.13 days per 10 years). Furthermore, on an annual scale, the minimum rate is recorded at the Casa Anfa station (0.98 days per 10 years) and the maximum rate is recorded at the Ouarzazate station (4.08 days per 10 years). Thus, it emerges from the results associated with the two deciles that the sub-desert regions (ex. Ouarzazate) experience a higher trend towards an increase in hot days than that recorded in the coastal resorts (ex. Tangier and Safi) or in the mountains (ex. Ifrane). However, the greater metropolis of Casablanca records the highest maximum increase for the 50th decile.

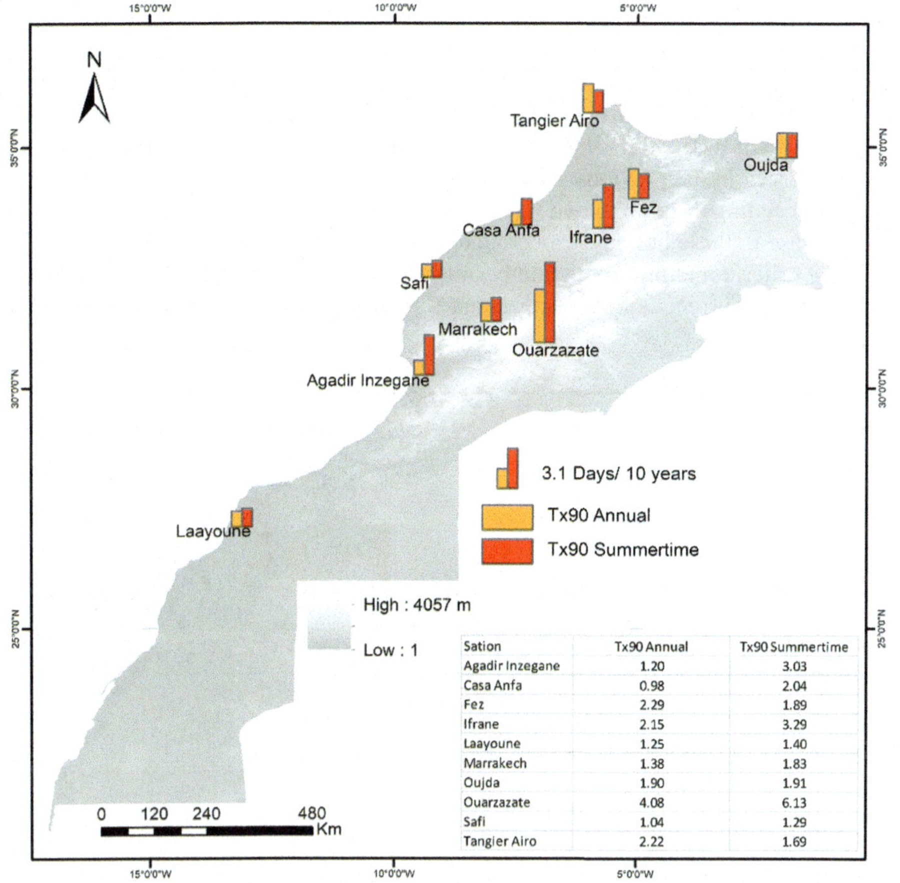

Sation	Tx90 Annual	Tx90 Summertime
Agadir Inzegane	1.20	3.03
Casa Anfa	0.98	2.04
Fez	2.29	1.89
Ifrane	2.15	3.29
Laayoune	1.25	1.40
Marrakech	1.38	1.83
Oujda	1.90	1.91
Ouarzazate	4.08	6.13
Safi	1.04	1.29
Tangier Airo	2.22	1.69

Fig. 2.3 The rate of increase in hot days per decade according to the Tx90th index (orange color shows the annual scale, red color shows the summer season (JJA))

These results suggest that the increase in hot days could have an impact on the overcoming of the record daily maximum temperatures. To examine this hypothesis, we study in the following approach the absolute records of Tmax and the advent of heat waves based on data from the reference period 1980–2019 in comparison with the records of June and July 2022.

Absolute Maximum Temperature Records

The absolute records for maximum temperature were extracted from daily temperatures for the reference period 1980–2019. These reference record values were compared with the maximum daily temperatures recorded during the months of

June and July of the year 2022, during which the country experienced heat waves over the majority of its territory. According to the results obtained, we observe that the record reference values were not exceeded during the months of summer 2022 (Fig. 2.4). However, all stations recorded their maximum record values during the last two decades of the period 1980–2019. This may confirm the trend of increasing hot days observed in the previous section, because there is an increase in the chance of values exceeding the 90th in two ways: first, hot days can last much longer; secondly the increase in these hot days can contain extreme or even record values. Nevertheless, the criteria relating to the number of hot days and record temperatures do not express the occurrence of heat waves that we will develop in the next section.

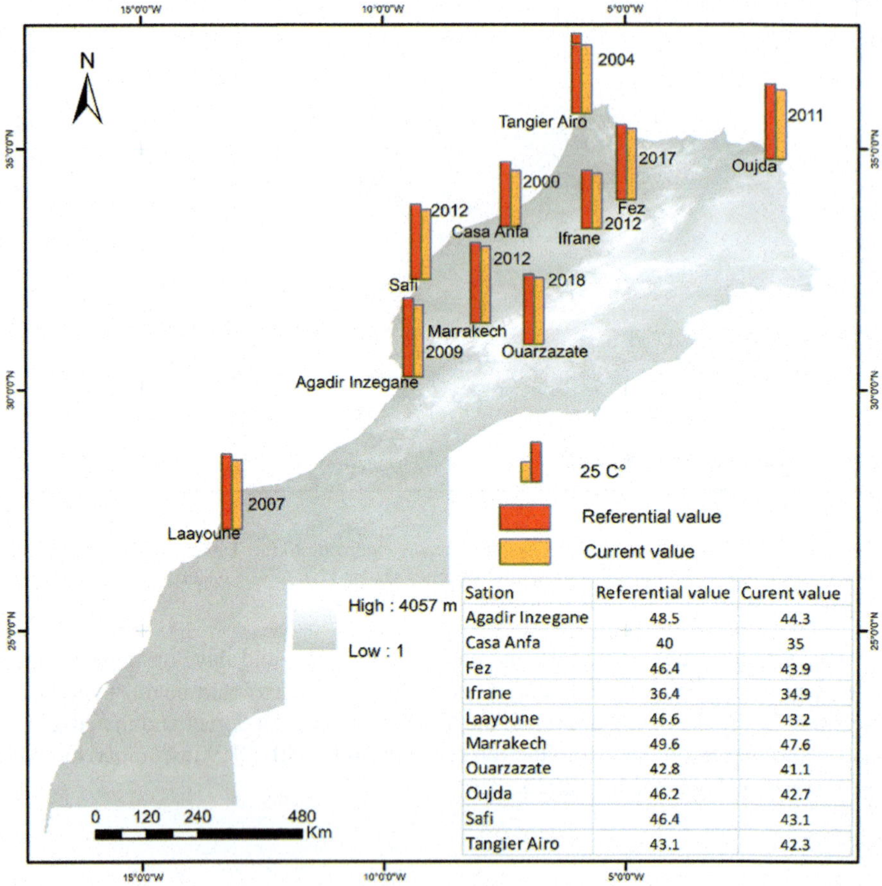

Sation	Referential value	Curent value
Agadir Inzegane	48.5	44.3
Casa Anfa	40	35
Fez	46.4	43.9
Ifrane	36.4	34.9
Laayoune	46.6	43.2
Marrakech	49.6	47.6
Ouarzazate	42.8	41.1
Oujda	46.2	42.7
Safi	46.4	43.1
Tangier Airo	43.1	42.3

Fig. 2.4 The absolute maximum values (the red color is the reference value of the base period 1980–2019; the orange color represents the current value of June and July 2022)

Heat Waves in Morocco

The Maximum Annual Number of Heat Waves

The maximum annual number of heat waves recorded during the reference period and the months of June and July 2022 show that four stations among all the stations studied recorded the same number calculated during the reference period (Fig. 2.5); these are the Tangier-Airo, Fez, Oujda and Marrakech station. This maximum annual number increased during the second decade of the 1980–2019 reference period.

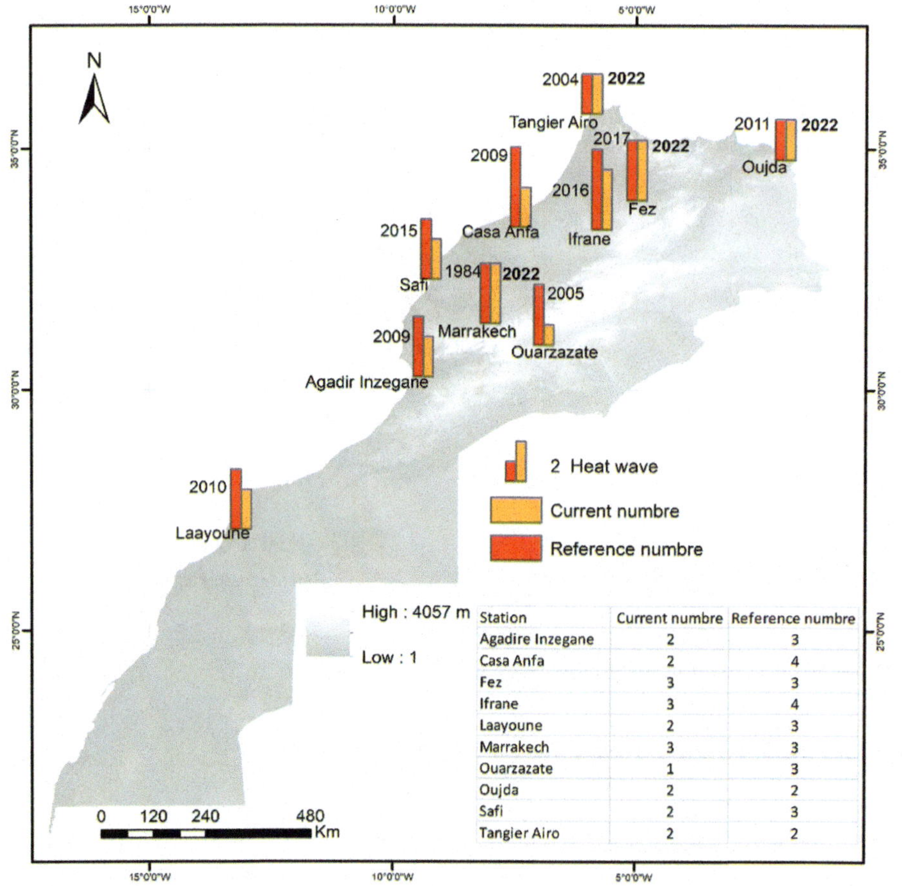

Station	Current numbre	Reference numbre
Agadire Inzegane	2	3
Casa Anfa	2	4
Fez	3	3
Ifrane	3	4
Laayoune	2	3
Marrakech	3	3
Ouarzazate	1	3
Oujda	2	2
Safi	2	3
Tangier Airo	2	2

Fig. 2.5 The maximum annual number of heatwaves during the reference period (red color) in comparison with the record for the months of June and July 2022 (orange). The indication of two dates for the same station indicates that the maximum number recorded in the reference period and in the months of June and July 2022

The Longest Maximum Duration of Heat Waves

The longest maximum duration of heat waves recorded in June and July 2022 exceeded the reference period in two out of 10 stations (Fig. 2.6). These are Casa-Anfa and Fez stations. The highest record was recorded at the Fez station during the summer of 2022 with a duration of 10 days compared to 8 during the reference period. For Casa-Anfa, this record went from 6 to 8 days.

The results confirm the increase in maximum temperatures over the last decades in Morocco (Khomsi et al. 2015; Filahi et al. 2016; Kamal et al. 2018; Driouech et al. 2020; Hadri et al. 2021; Zittis et al. 2021; Balhane et al. 2022; Hammoudy et al.

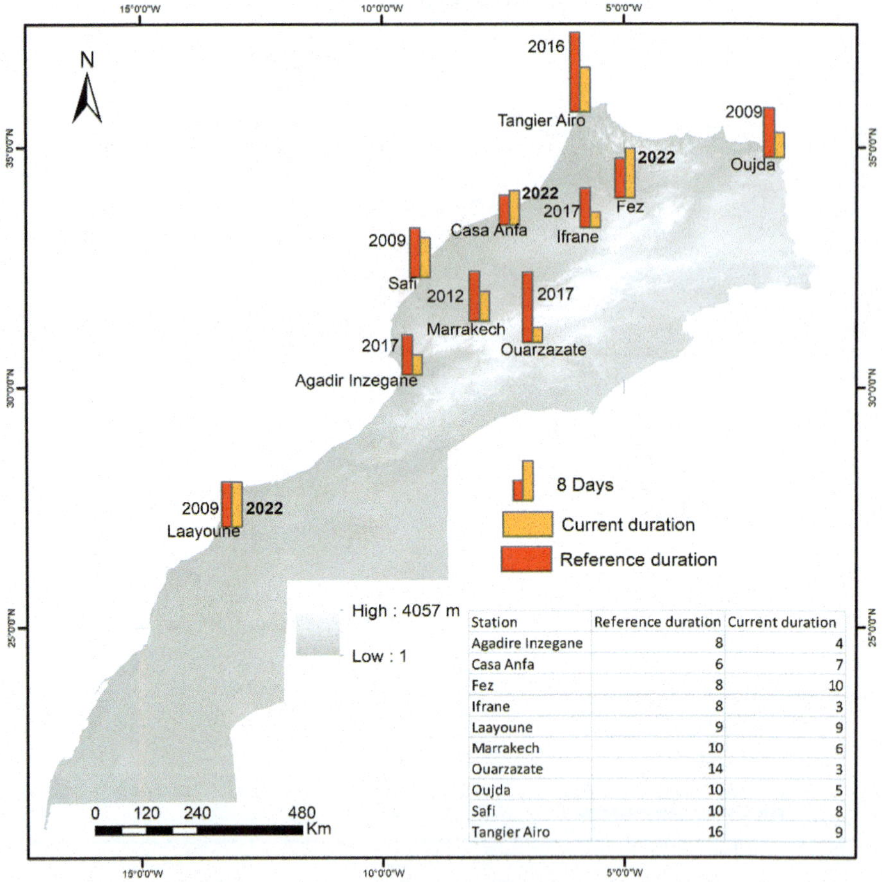

Station	Reference duration	Current duration
Agadire Inzegane	8	4
Casa Anfa	6	7
Fez	8	10
Ifrane	8	3
Laayoune	9	9
Marrakech	10	6
Ouarzazate	14	3
Oujda	10	5
Safi	10	8
Tangier Airo	16	9

Fig. 2.6 Comparison of the longest heat wave duration recorded during the reference period 1980–2019 (red color) with the longest heat wave during June and July 2022 (orange color). Years indicate referential or current records. The indication of two dates for the same station indicates that the maximum duration recorded in the reference period and in the months June and July 2022

2022). This increase stipulates an increase in the frequency of heat waves as well as an increase in their duration.

Conclusions

In recent years, Morocco, as for all Mediterranean countries, has experienced more and more the passage of summer heat waves which constitute a threat to water resources and human health. The summer of 2022 was marked by a repetitive passage of heat waves which were widely publicized. They were considered exceptional and attributed to climatic changes due to global warming.

This meteorological situation in the summer of 2022 led us to verify the exceptionality of heat waves by comparing them with a climatic reference period of 1980–2019. This is the only approach that allows us to verify the exceptionality of heat waves. To do this, a heat wave was defined when daily maximum temperatures exceed the 90th percentile for 3 consecutive days.

In a second step, the results of this study show that the assertion relating to the exceptionality of the heat waves of summer 2022 in comparison with the reference period is not fully verified. Indeed, the maximum temperatures and extreme values recorded during the heat waves of summer 2022 did not exceed the record values recorded during the reference climatic period. Thus, the extreme temperatures of summer 2022 cannot be considered exceptional. On the other hand, we have observed, for certain stations, during the summer of 2022, the increase in duration and number of heat waves.

The trend towards an increasingly hot summer with more frequent and longer duration heat waves could be the most dominant weather pattern over the next few years. It would be associated with a positive trend of hot and dry circulation types of Saharan origin. To validate this last hypothesis, we will address in our next work the link between heat waves and the types of atmospheric circulation.

Acknowledgements I would like to thank all officials of the Moroccan National Meteorological Directorate for their assistance in obtaining the approved data for this study.

References

Awasthi A, Vishwakarma K, Pattnayak KC (2022) Retrospection of heatwave and heat index. Theoret Appl Climatol 147(1–2):589–604. https://doi.org/10.1007/s00704-021-03854-z

Balhane S, Driouech F, Chafki O, Manzanas R, Chehbouni A, Moufouma-Okia W (2022) Changes in mean and extreme temperature and precipitation events from different weighted multi-model ensembles over the northern half of Morocco. Clim Dyn 58(1–2):389–404. https://doi.org/10.1007/s00382-021-05910-w

Basarin B, Lukić T, Matzarakis A (2020) Review of biometeorology of heatwaves and warm extremes in Europe. Atmosphere 11(12):1276. https://doi.org/10.3390/atmos11121276

Born K, Christoph M, Fink AH, Knippertz P, Paeth H, Speth P (2008) Moroccan climate in the present and future: combined view from observational data and regional climate scenarios. In: Climatic changes and water resources in the Middle East and North Africa, pp 29–45. https://doi.org/10.1007/978-3-540-85047-2_4

Cheng J, Xu Z, Bambrick H, Su H, Tong S, Hu W (2019) Impacts of heat, cold, and temperature variability on mortality in Australia, 2000–2009. Sci Total Environ 651:2558–2565. https://doi.org/10.1016/j.scitotenv.2018.10.186

Driouech F, ElRhaz K, Moufouma-Okia W, Arjdal K, Balhane S (2020) Assessing future changes of climate extreme events in the CORDEX-MENA region using regional climate model ALADIN-climate. Earth Syst Environ 4(3):477–492. https://doi.org/10.1007/s41748-020-00169-3

Filahi S, Tanarhte M, Mouhir L, El Morhit M, Tramblay Y (2016) Trends in indices of daily temperature and precipitations extremes in Morocco. Theoret Appl Climatol 124:959–972. https://doi.org/10.1007/s00704-015-1472-4

Founda D, Santamouris M (2017) Synergies between Urban Heat Island and Heat Waves in Athens (Greece), during an extremely hot summer (2012). Sci Rep 7(1):10973. https://doi.org/10.1038/s41598-017-11407-6

Founda D, Katavoutas G, Pierros F, Mihalopoulos N (2022) Centennial changes in heat waves characteristics in Athens (Greece) from multiple definitions based on climatic and bioclimatic indices. Glob Planet Change 212:103807. https://doi.org/10.1016/j.gloplacha.2022.103807

Hadri A, Saidi MEM, Saouabe T, El Alaoui El Fels A (2021) Temporal trends in extreme temperature and precipitation events in an arid area: case of Chichaoua Mejjate region (Morocco). J Water Clim Change 12(3):895–915. https://doi.org/10.2166/wcc.2020.234

Hammoudy W, Ilmen R, Sinan M (2022) Impact du changement climatique sur les extrêmes de températures au Maroc. In: 35ème colloque annuel de l'Association Internationale de Climatologie: Le changement climatique, les risques et l'adaptation, 6 pp. hal-03826000. https://hal.science/hal-03826000

Henri DH, Kluge P (2022) Statement—climate change is already killing us, but strong action now can prevent more deaths. World Health Organization

IPCC (2019) Climate change and land: an IPCC special report on climate change, desertification, land degradation, sustainable land management, food security, and greenhouse gas fluxes in terrestrial ecosystems. In: Shukla PR, Skea J, Calvo Buendia E, Masson-Delmotte V, Pörtner H-O, Roberts DC, Zhai P, Slade R, Connors S, van Diemen R, Ferrat M, Haughey E, Luz S, Neogi S, Pathak M, Petzold J, Portugal Pereira J, Vyas P, Huntley E, Kissick K, Belkacemi M, Malley J (eds). In press

Jiang J, Liu Y, Mao J, Wu G (2023) Extreme heatwave over Eastern China in summer 2022: the role of three oceans and local soil moisture feedback. Environ Res Lett 18(4):044025. https://doi.org/10.1088/1748-9326/acc5fb

Kamal AO, Ismail K, Abdelkrim A, Atika K, Driss E, Ezzahra EF, Zakaria A, Fatima N, Nadia N (2018) Climate change trend observations in Morocco: case study of Beni Mellal-Khenifra and Darâa-Tafilalt regions. J Geosci Environ Prot 6(07):34. https://doi.org/10.4236/gep.2018.67003

Khomsi K, Mahe G, Tramblay Y, Sinan M, Snoussi M (2015) Trends in rainfall and temperature extremes in Morocco. Nat Hazards Earth Syst Sci Discuss 3(2)

Kueh MT, Lin CY, Chuang YJ, Sheng YF, Chien YY (2017) Climate variability of heat waves and their associated diurnal temperature range variations in Taiwan. Environ Res Lett 12(7):074017. https://doi.org/10.1088/1748-9326/aa70d9

Lu R, Xu K, Chen R, Chen W, Li F, Lv C (2023) Heat waves in summer 2022 and increasing concern regarding heat waves in general. Atmos Ocean Sci Lett 16:100290. https://doi.org/10.1016/j.aosl.2022.100290

McGregor GR, Bessmoulin P, Ebi K, Menne B (2015) Heatwaves and health: guidance on warning-system development. World Meteorol Org World Health Org 1142:96p

Nairn J, Fawcett R (2011) Defining heatwaves: heatwave defined as a heat-impact event servicing all. Europe, pp 220, 224. https://www.cawcr.gov.au/technical-reports/CTR_060.pdf

Parente J, Pereira MG, Amraoui M, Fischer EM (2018) Heat waves in Portugal: current regime, changes in future climate and impacts on extreme wildfires. Sci Total Environ 631:534–549. https://doi.org/10.1016/j.scitotenv.2018.03.044

Perkins SE (2015) A review on the scientific understanding of heatwaves—their measurement, driving mechanisms, and changes at the global scale. Atmos Res 164:242–267. https://doi.org/10.1016/j.atmosres.2015.05.014

Perkins SE, Alexander LV (2013) On the measurement of heat waves. J Clim 26(13):4500–4517. https://doi.org/10.1175/JCLI-D-12-00383.1

Peterson T, Folland C, Gruza G, Hogg W, Mokssit A, Plummer N (2001) Report on the activities of the working group on climate change detection and related rapporteurs. World Meteorological Organization, Geneva, p 143

Sen PK (1968) Estimates of the regression coefficient based on Kendall's tau. J Am Stat Assoc 63(324):1379–1389. https://doi.org/10.1080/01621459.1968.10480934

Serrano-Notivoli R, Lemus-Canovas M, Barrao S, Sarricolea P, Meseguer-Ruiz O, Tejedor E (2022) Heat and cold waves in mainland Spain: origins, characteristics, and trends. Weather Clim Extrem 37:100471. https://doi.org/10.1016/j.wace.2022.100471

Stillman JH (2019) Heat waves, the new normal: summertime temperature extremes will impact animals, ecosystems, and human communities. Physiology 34(2):86–100. https://doi.org/10.1152/physiol.00040.2018

Tobías A, Royé D, Iñiguez C (2023) Heat-attributable mortality in the summer of 2022 in Spain. Epidemiology 34(2):e5–e6. https://doi.org/10.1097/EDE.0000000000001583

Tomczyk AM, Sulikowska A (2018) Heat waves in lowland Germany and their circulation-related conditions. Meteorol Atmos Phys 130(5):499–515. https://doi.org/10.1007/s00703-017-0549-2

Wie J, Moon BK, Hyun YK, Lee J (2021) Impact of local atmospheric circulation and sea surface temperature of the East Sea (Sea of Japan) on heat waves over the Korean Peninsula. Theoret Appl Climatol 144:431–446. https://doi.org/10.1007/s00704-021-03546-8

Winslow LA, Read JS, Hansen GJ, Hanson PC (2015) Small lakes show muted climate change signal in deepwater temperatures. Geophys Res Lett 42(2):355–361. https://doi.org/10.1002/2014GL062325

Yule EL, Hegerl G, Schurer A, Hawkins E (2023) Using early extremes to place the 2022 UK heat waves into historical context. Atmos Sci Lett e1159. https://doi.org/10.1002/asl.1159

Zhang X, Hegerl G, Zwiers FW, Kenyon J (2005) Avoiding inhomogeneity in percentile-based indices of temperature extremes. J Clim 18(11):1641–1651. https://doi.org/10.1175/JCLI3366.1

Zhang M, Yang X, Cleverly J, Huete A, Zhang H, Yu Q (2022) Heat wave tracker: a multi-method, multi-source heat wave measurement toolkit based on Google Earth Engine. Environ Model Softw 147:105255. https://doi.org/10.1016/j.envsoft.2021.105255

Zhang D, Chen L, Yuan Y, Zuo J, Ke Z (2023) Why was the heat wave in the Yangtze River valley abnormally intensified in late summer 2022? Environ Res Lett 18(3):034014. https://doi.org/10.1088/1748-9326/acba30

Zittis G, Hadjinicolaou P, Almazroui M, Bucchignani E, Driouech F, El Rhaz K, Kurnaz L, Nikulin G, Ntoumos A, Ozturk T, Proestos Y, Stenchikov G, Zaaboul R, Lelieveld J (2021) Business-as-usual will lead to super and ultra-extreme heatwaves in the Middle East and North Africa. npj Clim Atmos Sci 4(1):20. https://doi.org/10.1038/s41612-021-00178-7

Chapter 3
Climate Change Projection and Numerical Climate Modeling in the Rheris Watershed (Region Draa Tafilalet—Morocco)

Abdelmoumen Benchattou and Mohamed El Ghachi

Abstract Climate change is a widely recognized reality today in the scientific community. During the three decades, scientific researchers in Morocco spoke of climate variability, since the Rio Summit in 1992, we speak rather of climate change. This subject, of a transversal nature, interests many actors, coming from several climatological, economic, political, sociological fields, ... Each discipline has its own visions, objectives, means, problems etc. ... In the context of climate change, the Moroccan climate is characterized by a strong spatial and temporal irregularity. The Rheris basin is located in the South-East of Morocco, it covers an area of 12,702 km^2, it is characterized by a semi-arid climate. At the level of the watershed of the wadi Rheris, this climatic irregularity favors the occurrence of hydrological extremes (floods and floods), these hydrological risks have constituted in recent years a major risk in our study area. This scientific contribution has allowed us to focus on the recent, current and future climate of the Rheris watershed, based on optimistic and pessimistic scenarios of GEIEC to make the projection of future climate scenarios of the Rheris watershed and its impact on the occurrence of floods, which cause each time material damage and sometimes human losses.

Keywords Climate change · Rheris watershed · Climate variability · Hydrological risks · Impacts

A. Benchattou (✉) · M. El Ghachi
Laboratory for Landscape, Risk and Heritage Dynamics, Faculty of Arts and Humanities, Sultan Moulay Slimane University, Beni Mellal, Morocco
e-mail: benchattouabdelmomen@gmail.com

K. Kahime et al. (eds.), *Climate Change Effects and Sustainability Needs*, Springer Climate, https://doi.org/10.1007/978-3-031-59603-2_3

Introduction

In recent decades climate change has become a major issue in the modern world, climate change is a natural phenomenon, but can sometimes lead to dramatic consequences in some regions of the planet. Since the early 1980s, the southern Mediterranean region has experienced a series of dry years (Pachauri and Reisinger 2007). The southern and eastern Mediterranean countries are particularly vulnerable to climate variability because they are exposed to accelerated desertification, soil aridity and water scarcity in a context of population growth and increasing water demand. Due to its geographical position, Morocco has not escaped global climate change. It has an arid to semi-arid climate with a strong Mediterranean influence, which may be linked to local effects, a general trend, or in relation to broader phenomena (Study program "Climate change: impacts on Morocco and global adaptation options: 2019"). It is located in one of the regions where the impacts of climate change would be marked due to the amplification and increase in the frequency of extreme phenomena (floods). The Rheris basin is located in the South-East of Morocco, it covers an area of 12,702 km^2, it is characterized by a semi-arid climate. The interest in doing climate modeling in the Rheris watershed which is located in the South-East of Morocco in particular is undoubtedly related to the great importance of climate variability in the study area. Despite the arid and Saharan climate of the study area, the latter remains one of the regions of Morocco that experience floods and floods sometimes very devastating causing very costly damage, these hydrological extremes are generally due to heavy rains generalized that it is of short or long duration, the majority of floods occur in summer and autumn, This scientific contribution has allowed us to focus on the recent, current and future climate of the Rheris watershed, based on optimistic and pessimistic scenarios of GEIEC to make the projection of future climate scenarios of the Rheris watershed and its impact on the occurrence of floods, which cause each time material damage and sometimes human losses, and a feeling of insecurity among residents and tourists.

Materials and Methods

Presentation of the Study Area

The Rheris basin is part of the action area of the agency of the hydraulic basin of Oueds Guir-Ziz-Rheris (ABHGZR), it is located in the South-East of Morocco, it extends over an area of 12,702 km^2 the Rheris basin is limited to the North and East by the Ziz basin, to the North-West by the Oum-Rbia basin, to the West by the Draa basin and to the South by the Maider basin. The geology of the basin consists of Paleozoic (sandstone, quartzite and shale), Jurassic (predominantly limestone), Cretaceous (mainly limestone and sandstone) and plio-quaternary (alluvium, silt and dune sand). The climatic context is semi-arid, rainfall and temperature show

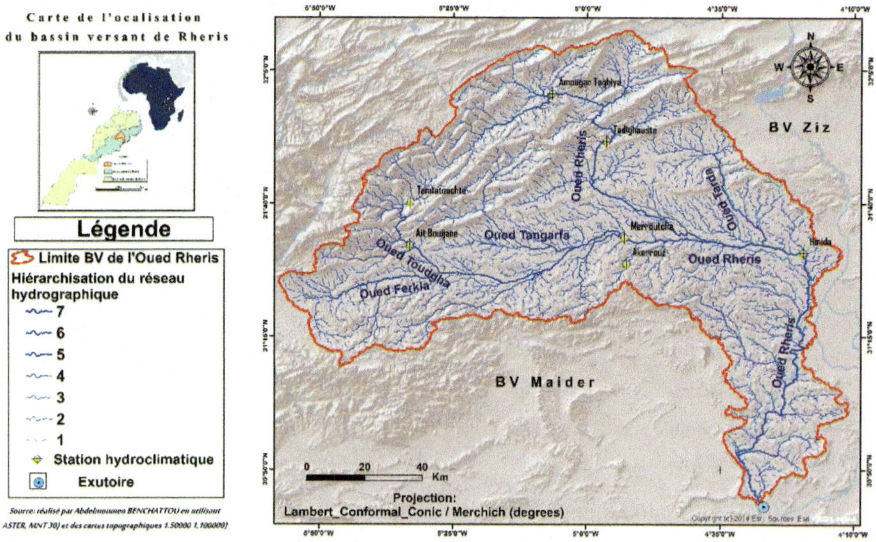

Fig. 3.1 Geographical location of the Rheris basin

irregularities in space and time, the study basin is characterized by rainfall concentrated in time and space, which favors the occurrence of floods and the damaging floods in the area (Fig. 3.1).

Methods and Data

According to the climate databases, the curve of the average temperature of the planet shows an increase in temperature, which gives a warming of the planet, and if we want to understand this curve and to interpret it, we need a climate model. In this scientific contribution, we used a climate model, which allowed us to focus on the recent, current and future climate of the Rheris watershed, based on optimistic and pessimistic scenarios of GEIEC to make the projection of future climate scenarios of the Rheris watershed and its impact on the occurrence of floods, which cause each time material damage and sometimes human losses, and a feeling of insecurity among residents and tourists (Fig. 3.2).

The KNMI Climate Change Atlas is a climate model for statistical analysis of climate data. It is developed by the Meteorological Institution of the Ministry of Infrastructure and Environment of the Royal Netherlands and contains more than 10 TB of climate data and dozens of analysis tools. It is part of the WMO Regional Climate Center at KNMI. Regarding the observed data used, a large part of the observation data is updated monthly, a part of the daily data is updated daily. Other data are updated as needed.

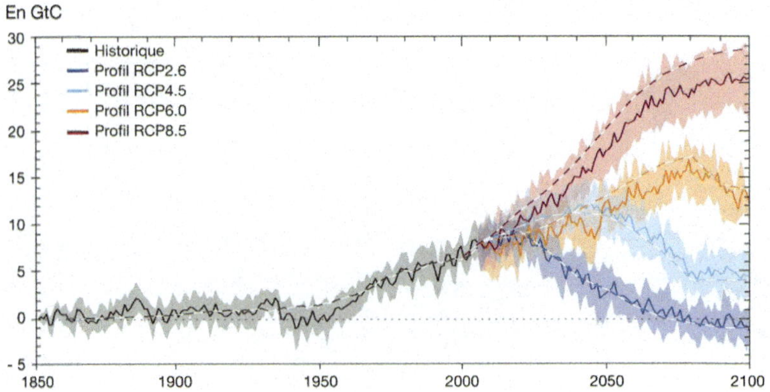

Fig. 3.2 Climate change scenarios (GEIEC)

The KNMI Climate Change Atlas model has developed four new scenarios for future climate change in the Netherlands around 2050 and 2085. The KNMI'14 climate scenarios provide a consistent picture of changes in 12 climate variables, including temperature, precipitation and sea level. Each scenario has a different storyline, depending for example on the amount of CO_2 emission. The four KNMI scenarios differ in the amount of global warming (moderate or warm) or possible changes in the air circulation pattern (low or high) (Fig. 3.3).

The KNMI'14 climate scenarios form the boundaries of likely future climate change. The scenarios will be used to map the impacts of climate change in order to assess the importance and urgency of climate adaptation measures. In this way, KNMI'14 climate scenarios help to make decisions for a safe and sustainable environment.

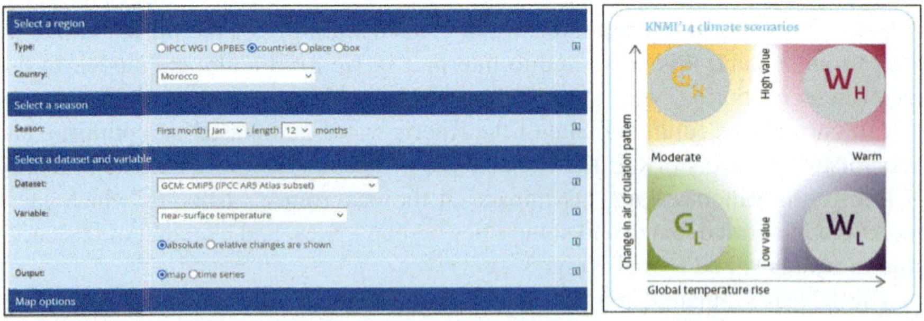

Fig. 3.3 The KNMI model interface and its different scenarios

Results and Discussion

Global Numerical Temperature Modelling on the AR5 CMIP5 Subset Model

According to the AR5 CMIP5 subset model, in all cases the planet will experience a warming in the future, all scenarios whether optimistic or pessimistic indicate that there is an increase in temperature, which will disrupt the climate system of the world and also will accelerate the appearance of a set of hydrological extremes (drought and floods) (Fig. 3.4).

Numerical Modeling of the Climatic Situation of Morocco on the AR5 CMIP5 Subset Model

As a North African country located between two climatic zones, temperate in the North and tropical in the South, Morocco is characterized by four types of climate: humid, sub-humid, semi-arid and arid. The data measured during the last decades attest to the progression of the semi-arid climate towards the north of the country. Climate projections for Morocco foresee a worsening of the climatic situation throughout the twenty-first century, as shown in Fig. 3.5 of temperature and precipitation compared to those of the period 1986–2020.

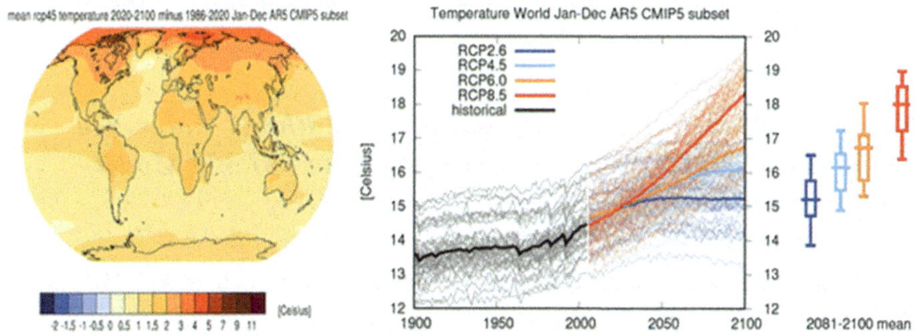

Fig. 3.4 Global numerical temperature modeling on the AR5 CMIP5 subset model

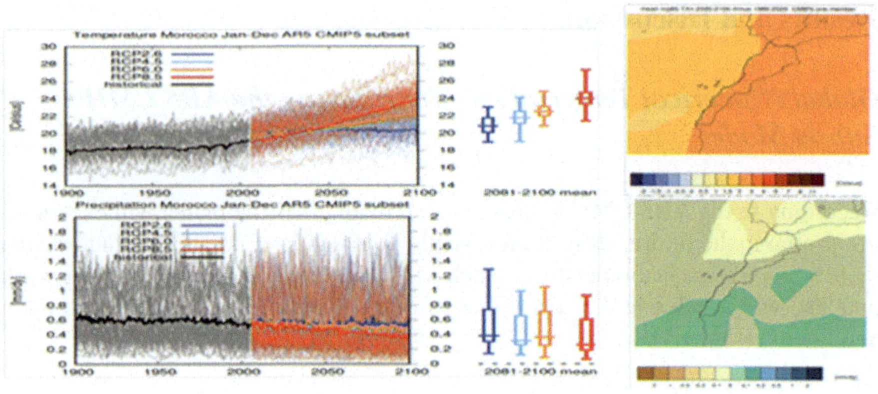

Fig. 3.5 Numerical modeling of the climate situation in Morocco on the AR5 CMIP5 subset model

Numerical Temperature Modeling of the Rheris Watershed on the AR5 CMIP5 Subset Model and on the SSP370 Model CNRM-CM6-1 Scenario

The Rheris watershed is located in the South-East of Morocco, it is characterized by a semi-arid climate. In the context of climate change, this watershed is experiencing a climatic disturbance, according to the AR5 CMIP5 subset model, in all cases the Rheris watershed will experience a warming in the future, all scenarios, whether optimistic or pessimistic, indicate that there is a temperature increase (Fig. 3.6).

Numerical Modeling of Precipitation in the Rheris Watershed on the AR5 CMIP5 Subset Model and on the SSP370 Model CNRM-CM6-1 Scenario

Regarding the projection of precipitation in our watershed, the AR5 CMIP5 subset model indicates that there is a decrease in precipitation in the future (2100) (Fig. 3.7).

According to the figures representing the spatial and temporal variation of rainfall in relation to the flows of the wadis of Rheris, we can observe that during the last years there is a decrease in rainfall and also we can observe that the majority of rainfall is summer rainfall in the form of thunderstorms, this type of rainfall that favors the appearance of floods and floods that cause each time material damage and sometimes human.

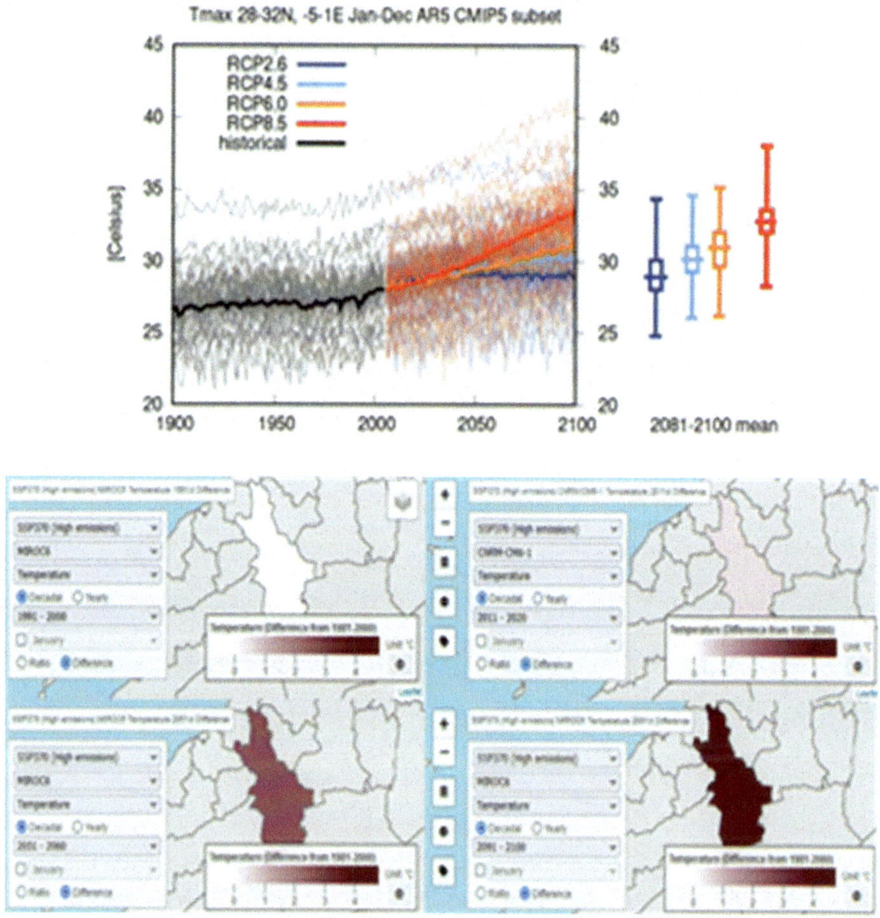

Fig. 3.6 Numerical temperature modeling of the Rheris watershed on the AR5 CMIP5 subset model and on the SSP370 Scenario CNRM-CM6-1

Potential Impacts of Climate Change in the Rheris Watershed

In the context of climate change and its impacts on the hydro-climatic system in the study area; during the last years the watershed of Rheris has experienced a climatic disruption, this climatic disruption represented in the form of rainfall concentrated in time and space, which directly promotes the occurrence of floods damaging, and an increase in temperature, Fig. 3.8 represent the temporal variability of rainfall in relation to the flows of wadis.

According to the figures representing the space–time variation of rainfall in relation to the flow of wadis of Rheris, we observe that during the last years there is a decrease in rainfall and also we note that the majority of precipitation are summer precipitation in the form of stormy rainfall, This type of precipitation favors the

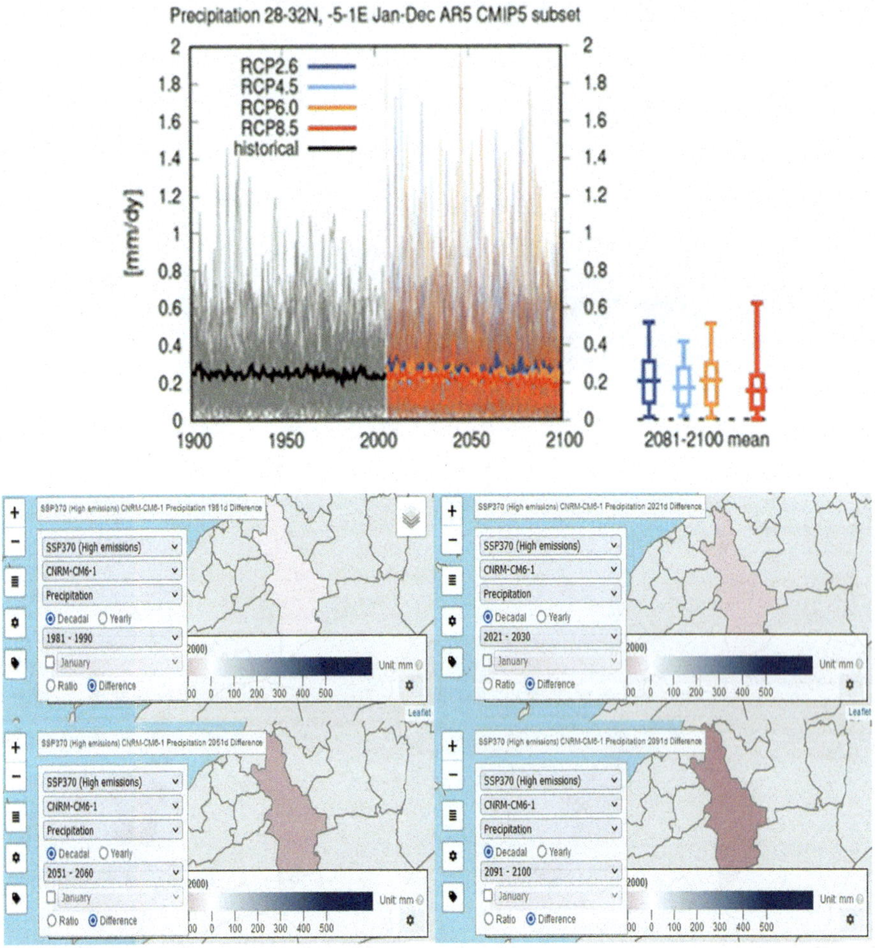

Fig. 3.7 Numerical modelling of precipitation in the Rheris watershed on the AR5 CMIP5 subset model and on the SSP370 model CNRM-CM6-1 scenario

appearance of floods and floods that cause each time material damage and sometimes human, in addition we find that the watershed of Rheris has experienced a climatic disturbance during the last years this area has a semi-arid climatic character knows snowy precipitation, as they shown in Fig. 3.9.

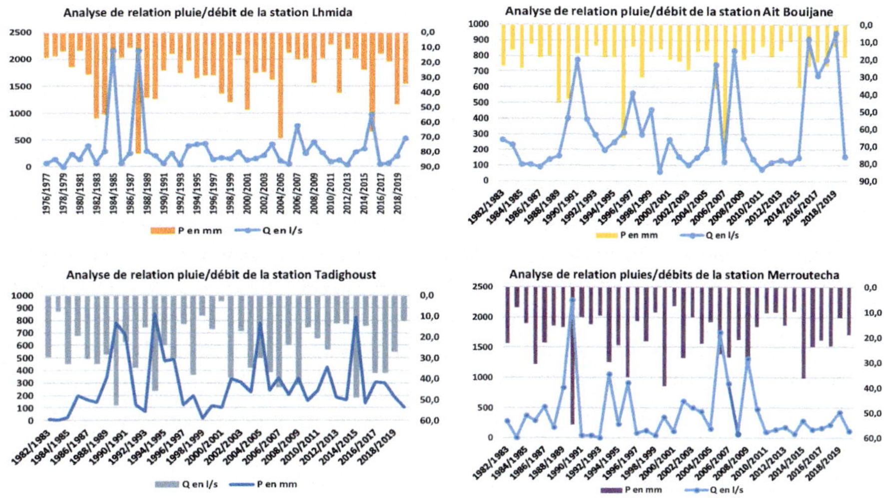

Fig. 3.8 Analysis of rainfall-flow relationship of the stations used

Fig. 3.9 Photo plate of the fieldwork 27/01/2022 in the upstream part of the Rheris BV

Conclusion

During the three decades, scientific researchers in Morocco were talking about climate variability, especially that of precipitation, in this scientific work we worked on the projection of climate change and numerical modeling of climate in the watershed of Rheris. In this study, the projections of climate variables (temperature and precipitation) were examined for the year 2100. All projections show an increase in temperature and decrease in precipitation in the study basin. The watershed of Rheris is already known a climatic disturbance, according to the analysis of the climatic data of the basin of Rheris, we noticed that our zone of study knew during the last years of the pluviometric precipitations concentrated in the time and in the space, this irregularity favors to the appearance of the hydrological extremes (the highs and the floods), these hydrological risks constituted a major risk for the national territory Especially for areas located at the foot of the mountain as the case of the basin of Rheris, the study basin is characterized by a very significant variability in rainfall, which negatively influences the operation of the wadi Rheris that can result in property damage and sometimes human losses, and cause a feeling of insecurity among residents and tourists.

Bibliography

Bastola S, Murphy C, Sweeney J (2011) The role of hydrological modeling uncertainties in climatic change impact assessments of Irish river catchment. Adv Water Resour 34(4):562–576

Brou YT (2009) Impacts of bioclimatic changes and forest land depletion in Ivorian peasantry: what solutions for sustainable agriculture in Cote d'Ivoire. Cuad Geogr 45:13–29

IPCC (2007) Intergovernmental Panel on Climate Change, climate change 2007. In: Contribution of working groups I, II and III to the fourth assessment report of the Intergovernmental Panel on Climate Change. IPCC, Geneva, 103 pp

Kont A, Jaagus J, Aunap R (2003) Climate change scenario and the effect of sea-level rise for Estonia. Glob Planet Change 36(1–2):1–15

Mackellar N, Christensen JH, Lucas-Picher P (2010) Climate models, projections and uncertainties in sub-Saharan Africa: an introduction for researchers and policy makers. CCDARE, 24 pp

Miller AI, Suppiah S, Duffer RB (2006) Climate change gains more from nuclear substitution than from conservation. Nucl Eng Des 236(16):1657–1667

Part II
Climate Change Effects on Ecosystems and Water Resources: Politics of Adaptation

Chapter 4
Genetic Control of *Bactrocera oleae* Gmelin (Diptera: Tephritidae) for Climate Change Mitigation

Tariq Nacimi, Driss Hmouni, and Yamna Ouguas

Abstract The olive tree (*Olea europaea*) is a highly prized resource in the Mediterranean basin due to its significant economic role. However, the olive tree is susceptible to various pests, which can exacerbate the effects of climate change. One major pest is the olive fly, *Bactrocera oleae* Gmelin (Diptera; Tephritidae), which inflicts significant damage on olive orchards by directly feeding on the drupes of the olives, leading to considerable yield reductions. Several control measures have been attempted to combat this pest, but some of these methods have limitations and can even have adverse effects on the environment and human health. Genetic control, based on the genetic resistance of olive cultivars, is considered an effective and environmentally friendly approach to combat *B. oleae*. In this study, we provide a description of the olive fly, including its distribution and life cycle, as well as an overview of the techniques used to control this pest. Furthermore, we delve into the topic of genetic control by presenting relevant studies conducted in this field. We discuss the factors involved in genetic control and the parameters used to determine the genetic resistance of olive tree cultivars against *B. oleae*.

Keywords *Bactrocera oleae* · *Olea europaea* · Control technics · Susceptibility · Genetic resistance

T. Nacimi (✉) · D. Hmouni
Department of Biology, Laboratory of Natural Resources and Sustainable Development, Faculty of Sciences of Kenitra, Ibn Tofail University, 133-14000 Kenitra, Morocco
e-mail: tariq.nacimi@uit.ac.ma

T. Nacimi · Y. Ouguas
Plant Protection Unit, Laboratory of Entomology, Regional Center for Agronomic Research of Marrakech, National Institute of Agronomic Research (INRA), 533-40000 Marrakech, Morocco

K. Kahime et al. (eds.), *Climate Change Effects and Sustainability Needs*, Springer Climate, https://doi.org/10.1007/978-3-031-59603-2_4

Introduction

With approximately 11.5 million ha of cultivated land worldwide, olive tree, *Olea europaea* L., is present in more than 50 countries mainly in the Mediterranean region (IOC 2022). The Mediterranean basin represents 91.83% of the world's olive crop (FAOSTAT 2019). The olive fly, *Bactrocera oleae* is considered a major pest of olive in the Mediterranean basin (Alfonso et al. 2014; Garantonakis et al. 2017). Females lay their eggs in the olive fruit and the emerged larvae feed on the mesocarp of the fruit (Pavlidi et al. 2017), significantly reducing the olive oil quality by deteriorating the olives and increasing the acidity of the oil produced (Zygouridis et al. 2009). The table olives become unpalatable (Daane and Johnson 2010), resulting in severe economic harm (Ponti et al. 2009). Several means of control have been tried namely, like the use of traps (Haniotakis 2005), biological control agents (Müller et al. 2019), sterile insect release (SIT), and chemical control (Vontas et al. 2011). Similarly, other control methods include the use of cultural practices, such as pruning, fertilization, plowing, and harvesting in appropriate time periods (Navrozidis et al. 2007), as well as the use of chemical repellents (Roessler 1989). These methods can contribute to the reduction of damage produced by the olive fly and to the protection of olive trees from infestation by this pest. The olive tree has a large varietal diversity, with more than 1200 cultivars spread around the world (Genç 2016). This high diversity could be used to select cultivars that show genetic resistance to *B. oleae*. The preference of olive flies for oviposition varied among olive cultivars. Studies carried out in this sense have shown that the susceptibility of certain cultivars is strongly related to the specific characteristics of the fruits of each cultivar, despite the fact that the cultivars studied are planted in the same site and under the same climatic conditions. Therefore, the evaluation of genetic resistance could exploit for the selection of olive cultivars less susceptible to the olive fly.

Overview of the Olive Fly (*Bactrocera oleae*)

The olive fly (*B. oleae*) is the main olive tree pest in the Mediterranean basin countries, causing important economic losses by degrading the quality and quantity of the olive crop.

The olive fly (*B. oleae*) is a species that belongs to the Brachycera suborder, the Diptera order, the Tephritoidea superfamily, and the Tephritidae family (Norrbom et al. 1999). The adult of this small species is about 5 mm long, with a wingspan of $\pm$ 10 mm.

Distribution of **B. oleae**

B. oleae Gmelin (Diptera: Tephritidae) is an insect found in many regions of the world. Across the Mediterranean basin, where 90% of the world's olive-growing areas are located, the olive fly is widespread in the Middle East, southern Europe and North Africa, regions in which 90% of the world's olive-growing areas are concentrated (Ponti et al. 2009). Similarly, this pest has been encountered in Asia, California (USA) and Mexico (Zygouridis et al. 2009). While, a study carried out by Müller et al. (2019) was demonstrated the potential establishment of *B. oleae* in olive growing regions where the species is absent.

The Life Cycle of **B. oleae**

The life cycle of the olive fly passes through four phases of development: egg, larva, pupa, and adult.

Egg: Adult females of *B. oleae* lay their eggs (usually only one egg per olive) inside the olives in the pulp (Fig. 4.1a, b), just under the epicarp, which allows the future larvae to have immediate access to nourishment after the egg emerges (Neuenschwander and Michelakis 1978, 1981). According to Zalom et al. (2009), under laboratory conditions, a single female can lay between 200 and 500 eggs over her lifetime, and about 10–40 eggs per day. However, in the wild, a female can lay around 12 eggs per day, and between 200 and 250 eggs over her lifetime (Ahmadi et al. 2018). Oviposition sites are simple to identify with the naked eye because a brown mark develops following oviposition (Fig. 4.1d). Olive fly eggs cannot survive at temperatures above 30 °C or below 32 °C (Tsitsipis 1977). In ideal circumstances, eggs can develop inside the mesocarp into larvae in just one day and make the galleries in the fruit.

Larva: According to their developmental stage, larvae are categorised as L1, L2, and L3 (Fig. 4.1c). Temperatures for larval development should range from 10 to 32 °C (Tsitsipis 1977). The *B. oleae* larvae may attain full maturity in 8 days, with an interval of 8–37 days, while ingesting 45–150 mg of olive pulp under ideal conditions (Neuenschwander and Michelakis 1979).

Olive fly larvae are monophagous as they feed exclusively on *O. europaea* species, on the other hand, adults of *B. oleae* species are polyphagous as they can use diversified sources for their nutrition. They obtain their nutrient from fruit exudates (Athar 2005; Ben-Yosef et al. 2014), plant nectar and pollen, bacteria, yeast, and insect honeydew (Daane and Johnson 2010).

Pupa: It is a transitional stage between the larval and adult state, also the conservation of the species during unfavorable periods, especially winter (Margier 2012). The pupa is elliptical, whitish yellow at the beginning of this stage and dark brown at the end

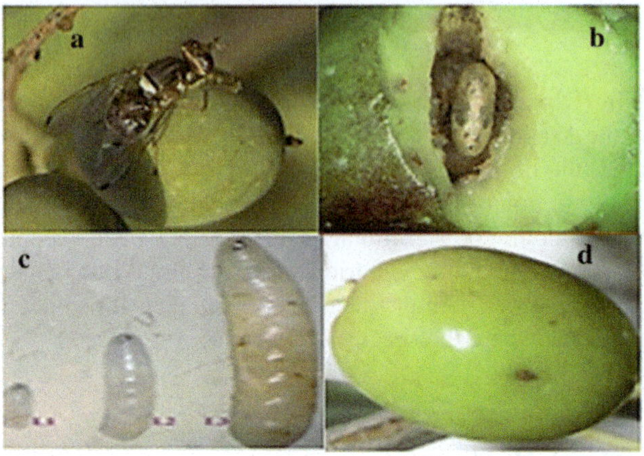

Fig. 4.1 An adult olive fly undergoing oviposition (**a**); olive fly egg (**b**); olive fly larva (**c**); oviposition site (**d**)

of this stage, measuring 5.1 mm in length and 2.1 mm in width (Genç and Nation 2008).

Adult: The adult is 4–5 mm long (same for the male and female). The head is orange with large blue-green faceted eyes and contains small antennae (Rice 2000). The legs are orange. The thorax is yellow-red, black-backed with four gray stripes ending in a yellow to white scutellum with several yellow-white spots on each side of the thorax. The wings are transparent with a characteristic black spot on the tips of the wings. The wing veins are slightly darkened.

Female olive flies can be distinguished from males by a dark abdominal base and the presence of an ovipositor for egg laying (Margier 2012).

Damage Caused by B. oleae

B. oleae is the most damaging pest for olive cultivation in the whole Mediterranean basin due to the direct destruction of the olive pulp caused by the development of the larvae, the premature fall of the infested fruits and the general reduction of the quality of the olive oil, which can lead to a loss of up to 40% of the production (Boccaccio and Petacchi 2009; Pereira et al. 2004).

Control Techniques Against B. oleae

To control the population of *B. oleae* and reduce the harmful effect on olive orchards several measures are used.

Cultivation Practices

The availability of water and the presence of stagnant water in the orchards favor the survival of the adults of the olive fly, especially during the hot periods of the year. Therefore, it is necessary to control the leakage of irrigation systems that can leave abundant water for the adults of this pest. Also, harvesting at the right time is much recommended to avoid the development of a large population of *B. oleae* adults in the orchard, minimizing damage and losses. After harvesting, it is highly recommended to turn the soil (plowing) in order to eliminate the pupae remaining in the soil. Also, the correct pruning, the exposure of the tree to the sun and balanced fertilization are important factors in the reinforcement of the resistance of the trees against this pest; Post-harvest sanitation plays a major role in the control of *B. oleae* to remove any remaining fruit in trees that may be infested and have damage in the next harvest (Yokoyama 2015). Thus, field cleaning of fallen fruits is necessary to reduce outbreaks for future generations (Ouguas 2021).

Mass Trapping

According to El-Sayed et al. (2006, 2009), mass trapping has become an alternative tool to the use of insecticides to control crop pest populations, especially Lepidoptera and Diptera. Trapping techniques are based on the principle of attracting insect pests (preferably females) to the devices to remove them from the rest of the population, either by keeping them in the device or by exposing them to toxic substances (Yasin et al. 2014). The attractant can be a pheromone or a food attractant with the possibility of adding an insecticide. Indeed, this type of control is considered an effective alternative in view of its capacity and the low cost they generate (Altolaguirre-Obrero et al. 2003).

Mass trapping consists of using plastic bottles (of a volume of 1–1.5 l) attached to the trees at man's length (preferably one bottle per tree) contain an attractive solution to the insects, composed of 5% mineral fertilizer (ammonium sulfate or ammonitrate) at a concentration of 50 g/l, perforated at the level of the upper two thirds by 3–4 holes of 3–5 mm in diameter. This trap should be checked every 10–15 days and renewed when the water evaporates (Ouguas 2021).

Biological Control of *B. oleae*

To limit the use of insecticides against the olive fly, the use of biological control is the most important environmentally friendly alternative (Hladnik 2017). In general, natural enemies like parasitoids, nematodes, fungus, bacteria, viruses, substances from a living origin, and other predators are the most often used biocontrol agents against insect pests (Lacey et al. 2001).

Parasitoids are biological control agents that are being studied for their important role in the control of the olive fruit fly. These natural enemies include some arthropods like native ants (*Formica aerata*) (Orsini et al. 2007), and *Psyttalia concolor* (Szépligeti) (Yokoyama et al. 2008).

Studies on fungi and their toxins have revealed a significant role in the management of olive flies. Konstantopoulou and Mazomenos (2005) demonstrated that fungal isolates obtained from *B. oleae*-infested olives showed significant mortality in the olive fruit fly batches tested. A similar study conducted by Yousef et al. (2013) to evaluate the effect of the fungus *Metarhizium brunneum* and its extracts on the olive fly revealed promoting results in controlling the density of adults of this pest.

Indeed, some vegetable essential oils have repellent capacities against the olive fruit fly (Giunti et al. 2022), while other essential oils have been shown to attract and destroy the olive fruit fly (Malheiro et al. 2015a). Similarly, a study published by Rizzo et al. (2020) showed that the treatment of olive flies with essential oils of *Pimpinella anisum* and *Trachyspermum ammi* caused significant mortality in this pest.

According to the results of research conducted by Torrini et al. (2017, 2020), the entomopathogenic nematodes *Steinernema feltiae* and *Heterorhabditis bacteriophora* can be a useful tool in the biological control of the olive fruit fly at various stages of its life cycle. Laboratory testing of olive fruit fly susceptibility to six commercially available entomopathogenic nematodes showed that *S. feltiae* potentially suppresses the third stage of larval development (Sirjani et al. 2009).

Despite the advances achieved in the study of biological control agents, there are still some barriers to their application in the field these include legal restrictions, the permissible amount of parasitism in the wild, and the enormous expenses associated with mass rearing (Ovruski and Schliserman 2012).

Chemical Control of *B. oleae*

Chemicals remain the most widely used means of controlling crop pests including fruit flies (Tephritids) (Roessler 1989). Chemical control has been the mainstay of the management of *B. oleae* since the middle of the last century. The most widely used plant protection products are organochlorines (OCLs), organophosphates (OPs), and carbamates (CARBs), following the more recent introduction of pyrethroids, spinosad and other compounds (Vontas et al. 2011). The application of insecticides in olive cultivation against *B. oleae* shows satisfactory results for farmers, as they cause significant mortality rates in the populations of this pest. However, excessive pesticide use can have negative effects on both human health and the environment, as well as the emergence of cross- and multiple-resistances in populations of the targeted insects (Sparks and Nauen 2015). The chemicals can negatively impact beneficial insects (Daane et al. 1990).

Genetic Control of *B. oleae*

In order to reduce the use of plant protection products and avoid these harmful impacts on humans and the environment, genetic control is an important tool that allows both effectively control the olive fly and reduce the damage on crops and the conservation of the environment and the health of consumers.

Numerous of studies suggest that olive cultivars differ in their susceptibility to olive fly (Rizzo and Caleca 2006; Daane and Johnson 2010; Malheiro et al. 2018). Indeed, various olive cultivars cultivated in the same region may be more or less susceptible to the olive fly (Daane and Johnson 2010). Other studies showed that different cultivars of olives have different oviposition preferences and larval performance (Gonçalves et al. 2012; Gümusay et al. 1990; Iannotta et al. 2007).

Genetic control may also represent an important tool to control *B. oleae* (Gmelin). The objective of this work is to characterize 21 olive cultivars with respect to their infestation by *B. oleae* and to discuss the correlation between the parameters involved in this infestation to contribute to the selection of cultivars that are more resistant to this pest.

Materials and Methods

Fields and Samplings

Every 15 days, from mid-July to the end of November, samples of 40 fruits were taken from two plants per cultivar. For cultivars with low production, only one tree was sampled. A total of 21 olive cultivars of various origins were sampled; 14 belong to the experimental area of Tassaout, located 70 km from Marrakech on the road to Fez, and seven belong to the experimental area of Saada, located 7 km from Marrakech on the road to Essaouira (Table 4.1). The ripeness index of the olive, the infestation rate and the intensity of the infestation by *B. oleae* were evaluated under a binocular magnifying glass when brought back to the laboratory.

Measured Parameters

The Ripeness Index

We determined the ripeness index or maturity index (IM) based on the assessment of the coloring of 100 olives taken randomly from each cultivar. These olives have been split into eight classes ranging from olives with intense green or dark green skin to olives with black skin and completely dark pulp (Fig. 4.2).

The olives ripeness index (RI) is calculated by the formula bellow:

Table 4.1 Studied olive cultivars and their geographic origin

Field	Olive cultivars	Geographic origin
Saada	*Picholine Marocaine* *Berri Meslal* *Ronde de la Ménara* *Meslala* *Menara* *Haouzia* *Tassaout* *Mechkat* *Dalia* *Baraka* *Agdal*	Morocco
Tassaout	*Arbequina* *Hojiblanca* *Picual* *Blanqueta* *Morisca* *Cornicabra* *Manzanilla de la Sevilla*	Spain
	Cerasuola *Pendolino*	Italy
	Koroneiki	Greece

Fig. 4.2 Grid for determining the maturity index (IOOC 2007, cited in Ornek et al. 2019)

$$RI = \sum ni \times i/100$$

(*i: the class; ni: the number of olives in the class; RI varies from 0 to 7*).

The fruit is classified 0 when the epidermis is intense green; 1 if the epidermis is yellowish green; 2 if the epidermis shows red colorations on less than half of the fruit; 3 if the epidermis is red or purple on more than half of the fruit; 4 if the epidermis is black and the pulp is white; 5 if the skin is black and less than half of the pulp is purple; 6 if the skin is black and more than half of the pulp is purple (without reaching the stone); 7 if the skin is black and the pulp is completely purple (up to the olive kernel).

$$RI = (a \times 0) + (b \times 1) + (c \times 2) + (d \times 3) + (e \times 4) + (f \times 5) + (g \times 6) + (h \times 7)/n$$

The letters (a, b, c, d, e, f, g, h) represent the number of fruits in each class and "n" is the number of observed fruits.

The Level of Olive Infestation

The infestation level is the ratio of the infested fruits to the total number of fruits sampled. We considered an olive infested by the olive fly when it shows feeding punctures or exit holes.

The Infestation Intensity of the Olives

From the 40 olives sampled from each plant, the average number of bites per olive was noted to give the infestation intensity.

Statistical Analysis

ANOVA at a significance level of 5% was used to compare the different cultivars studied with respect to the observed parameters and PCA (Principal Component Analysis) to characterize and classify the cultivars according to the investigated parameters. The data were analyzed by the software R.

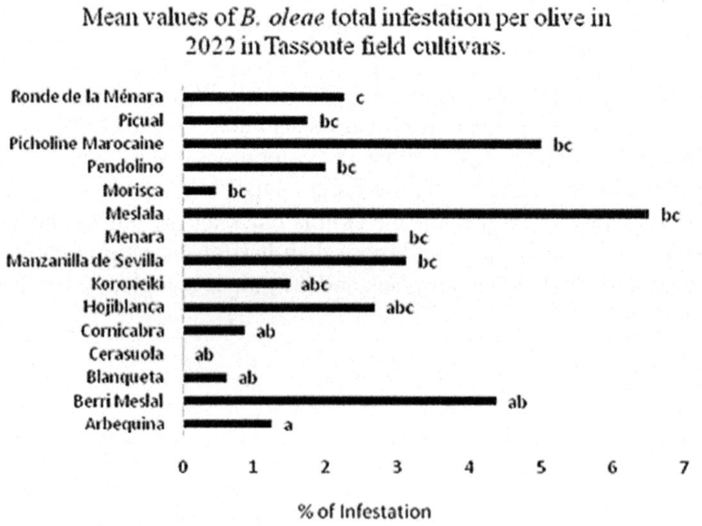

Fig. 4.3 Evaluation of the infestation by *B. oleae* on the olive cultivars of Tassaout field

Results and Discussion

Cultivars of Tassaout Field

The Level of Infestation by the Olive Fruit Fly

The results from the statistical analyze allowed the formation of five homogeneous groups. Then, for the olive cultivars of the experimental field of Tassaout, the results obtained from the statistical analyze concerning the infestation rate by *B. oleae* allowed the formation of five homogeneous groups. The same results showed that the highest infestation rate was recorded in the Meslala cultivar with 6.5, followed by the Berri Meslal cultivar (4.375) and Manzanilade Sevilla (3.125). An average infestation rate was registered for the Menara cultivar, Hojiblanca and Ronde de la Menara whit 3.00, 2.678 and 2.25 respectively. The lowest intensity of infestation was recorded in Morisca (0.468), Blanqueta (0.625), Cornicabra (0.875), Pical (1.75), Koroneiki (1.5) and Arbequina (1.25). Among all the cultivars, the Cerasuola cultivar hasn't show any infestation (Fig. 4.3).

Infestation Intensity of the Olives

The results showed the presence of a very highly significant difference between the infestation intensity means of the cultivars studied ($p < 1\%$ at 14 degree of freedom). Then, a highly significant effect of cultivar on infestation intensity ($p < 1\%$). The

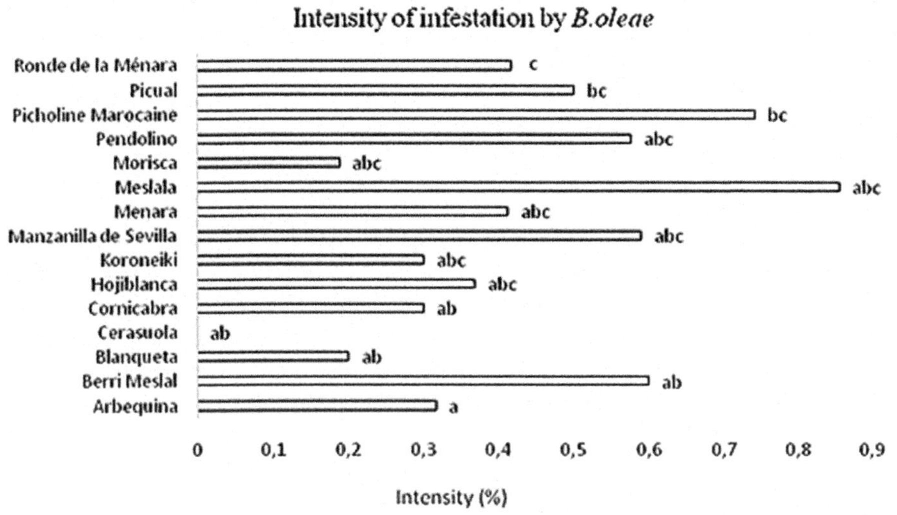

Fig. 4.4 Evaluation of the intensity of infestation by *B. oleae* in the olive cultivars of Tassaout field

Tukey's test led to the formation of five homogeneous groups: a, ab, abc, bc and c. It should be noted that the cultivar Arbequina is separated in one group and the other cultivars in four overlapping groups (Fig. 4.4).

The results obtained showed that the highest infestation intensity was recorded in the cultivar Meslala (0.93), followed by the cultivar Berri Meslal (0.755), Pendolino (0.75) and Marocain Picholine (0.74). The lowest infestation intensity was recorded in the cultivar Blanquita (0.2). The other cultivars had intermediate infestation intensities.

The Correlation Between the Maturity Index and the Infestation by *B. oleae*

Regarding the correlation between the infestation by *B. oleae* and the ripeness index of olives of the cultivars studied in the experimental field of Tassaout, the results obtained revealed that in addition to the cultivar parameter, the stage of ripening of olives appears as a preponderant factor influencing the laying of olive fly and therefore the intensity of infestation (Fig. 4.5).

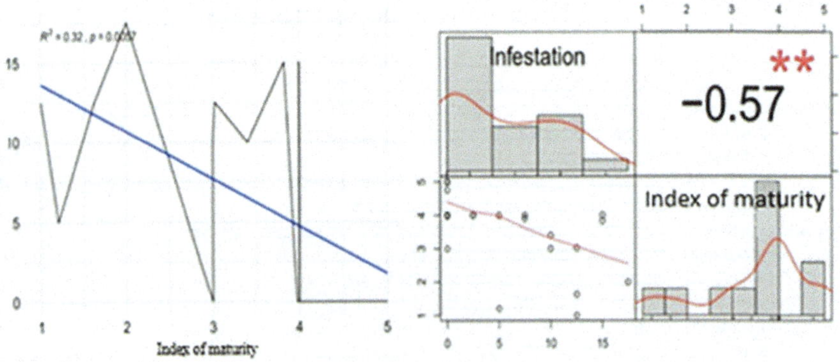

Fig. 4.5 The correlation between the infestation and the maturity index of olives of the cultivars studied in the Tassaout field

Cultivars of the Saada Field

Level of Infestation by *B. oleae*

The results obtained showed that the effect of cultivar was not significant ($p > 0.5$). Indeed, these results revealed that the highest infestation rate was recorded in the cultivar Haouzia (0. 781), then Agdal and Dalia (0.625), Menara (0.50), Mechkat (0.25), while the lowest infestation rate by *B. oleae* was recorded in the cultivar Baraka (0.156) and Tassaout (0.052) (Fig. 4.6).

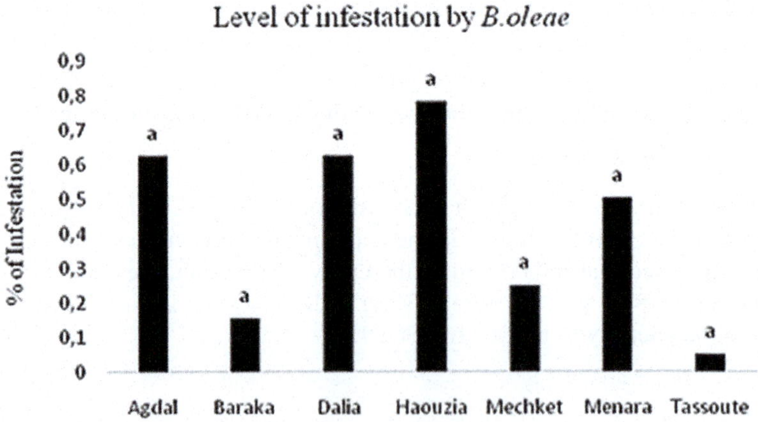

Fig. 4.6 The level of infestation by *B. oleae* on the olive cultivars of Saada field

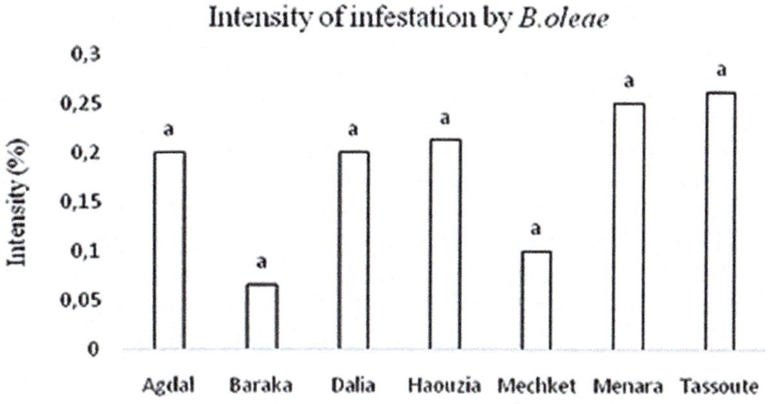

Fig. 4.7 Intensity of infestation by *B. oleae* on the olive cultivars of Saada field

Infestation Intensity of the Olive

The results obtained revealed that the highest infestation intensity was recorded in the Tassaout cultivar (0.2625), then Menara (0.250), Haouzia (0.212), Dalia and Agdal (0.20), Mechkat (0.10), while the lowest infestation intensity by *B. oleae* was recorded in the Baraka cultivar with 0.065 (Fig. 4.7).

The Correlation Between the Maturity Index and the Infestation by *B. oleae*

The results obtained revealed a negative correlation in the cultivars studied in the experimental field of Saada, but it does not seem to be significant; only 32% of all the cultivars meet this correlation (Fig. 4.8).

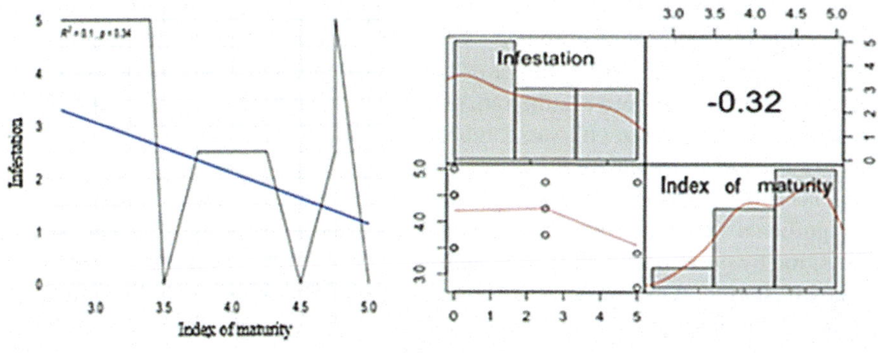

Fig. 4.8 The correlation between the infestation and the maturity index of olives of the cultivars of Saada field

The results of this study showed that the cultivars tested do not have the same resistance to the olive fruit fly, despite the fact that the cultivars studied belong to the same geographical site and are subject to the same environmental conditions. A study conducted by Malheiro et al. (2018) within an integrated production olive grove located in Paradela in the Trás-Os Montes region (northeast Portugal) focused on three of the main cultivars of the Trás-os-Montes region, Cobrançosa, Madural and Verdeal Transmontana. The results obtained regarding the level of infestation indicated that both cultivars Madural and Verdeal Transmontana are more susceptible to olive fly infestation than the cultivar Cobrançosa. The results obtained showed significant differences between the two cultivars Verdeal Transmontana and Cobrançosa, with an infestation rate of 24.5% observed with the cultivar Verdeal Transmontana, followed by Madural (16.0%), and finally the cultivar Cobrançosa (8.5%). Similar results were obtained by Gonçalves et al. (2012) before.

Medjkouh et al. (2018) worked on the susceptibility of olive eight olive cultivars showed that the Rougette de Mitidja is the cultivar most attacked by the olive fly with a rate of 65.33%. The cultivars Souidi and Chemlal revealed the lowest percentage of infestation (21%).

The study conducted by Rizzo and Caleca (2006) during the period 2002–2005 on 18 cultivars from Castelvetrano (province of Trapani, Sicily), revealed that, among the cultivars tested, Nocellara del Belice was found to be the most susceptible to olive fly attack. While, Giarraffa and Pizzo di corvo average sensitivity to attacks by *B. oleae*. The Minuta cultivar showed the lowest sensitivity.

The maturity index has an important effect on the rate and intensity of olive fly infestation. Indeed, the colour of the olives is one of the decisive factors in the oviposition preference of female olive flies (Yokoyama 2012). In general, the olive fly prefers greener fruits compared to ripe fruits (Katsoyannos and Kouloussis 2001; Malheiro et al. 2018). According to Vignesh and Chandrasekaran study (2020), fruits in the transitional stage from worm to yellow color are more attacked by the olive fly than in other ripening stages. Whereas, dark coloration of ripe fruit could disrupt drupe recognition by female olive fruit flies, reducing infestation levels in olives that reach advanced stages of ripening (Iannotta and Scalercio 2012).

Overall, the current study indicates that among the foreign cultivars of the Tassaout field, a highly significant difference between the average rate of infestation by *B. oleae*, and a highly significant effect of cultivar on infestation intensity. Therefore, a negative correlation between the rate of infestation and the ripening of the olives. Then, among the foreign cultivars studied, the Spanish cultivar Arbequine and the Italian cultivar Cerasuola are the least attacked by *B. oleae*.

Concerning the Moroccan cultivars of the Saada field, the results obtained showed no significant difference between the cultivars studied regarding the intensity of infestation by the *B. oleae*, and a non-significant effect of the cultivar on the intensity of the infestation. As a consequence, a negative correlation between the parameters examined, but it seems not significant. Among the Moroccan cultivars, it should be noted that for this year the Baraka and Mechket cultivars are the least attacked.

Conclusions

The considered study allowed us to classify the olive cultivars of Tassaout and Saada olive germplasm according to the infestation rate and intensity of olives. Also, the type of correlations between the rate of infestation and the index of maturity of olives was brought up in the two sites of the study. This first classification of its kind constitutes an interesting basis but it will not be able to assess the behavior of each cultivar with respect to the olive fly. To better characterize this infestation, it is recommended that the same study be conducted again in 2023 and 2024 to take into account interannual effects. It is also recommended that this study be extended to all olive cultivars of the Tassaout germplasm, which includes 591 olive cultivars of different geographic origins. This ultimate characterization could might be used in the classification and especially in the selection of additional olive tree genotypes that are more resistant to the olive fly, which genotypes could be used in a genetic improvement program through crosses or other selection procedures. Climate change may be mitigated by increasing the profile of olive fly tolerant cultivars.

References

Ahmadi M, Salehi B, Abd-Alla AMM, Babaie M (2018) Feasibility of using the radiation-based sterile insect technique (SIT) to control the olive fruit fly, *Bactrocera oleae* Gmelin (Diptera: Tephritidae) in Iran. Appl Radiat Isot 139:279–284

Alfonso I, Vacas S, Primo J (2014) Role of α-copaene in the susceptibility of olive fruits to *Bactrocera oleae* (Rossi). J Agric Food Chem 62:11976–11979

Altolaguirre-Obrero M, López-Pérez A, Caballero-Jiménez JA (2003) Estrategia alternativa al control de mosca del olivo *Bactrocera oleae* (Gmelin) mediante "trampa OLIPE". Ensayos en distintas zonas de la provincia de Córdoba. In: XI Simposium Científíco-Técnico Expoliva, Jaén, Spain, pp 14–16

Athar M (2005) Infestation of olive fruit fly, *Bactrocera oleae*, in California and taxonomy of its host trees. Agric Conspec Sci (ACS) 70(4):135–138

Ben-Yosef M, Pasternak Z, Jurkevitch E, Yuval B (2014) Symbiotic bacteria enable olive flies (*Bactrocera oleae*) to exploit intractable sources of nitrogen. J Evol Biol 27:2695–2705

Boccaccio L, Petacchi R (2009) Landscape effects on the complex of *Bactrocera oleae* parasitoids and implications for conservation biological control. BioControl 54:607–616

Daane KM, Johnson MW (2010) Olive fruit fly: managing an ancient pest in modern times. Annu Rev Entomol 55:151–169

Daane KM, Dahlsten DD, Dreistadt SH (1990) Effects of Mediterranean fruit fly malathion bait spray on the longevity and oviposition of parasitoids of linden and tuliptree aphids (Homoptera: Aphididae). Environ Entomol 19:1130–1134

El-Sayed AM, Suckling DM, Wearing CH, Byers JA (2006) Potential of mass trapping for long-term pest management and eradication of invasive species. J Econ Entomol 99(5):1550–1564

El-Sayed AM, Suckling DM, Byers JA, Jang EB, Wearing CH (2009) Potential of "lure and kill" in long-term pest management and eradication of invasive species. J Econ Entomol 102(3):815–835

FAOSTAT (2019) FAOSTAT statistics database. Food and Agriculture Organization of the United Nations (FAO), Rome. http://www.fao.org/faostat/en/#home

Garantonakis N, Varikou K, Birouraki A (2017) Parasitism of *Psyttalia concolor* (Hymenoptera: Braconidae) on *Bactrocera oleae* (Diptera: Tephritidae) infesting different olive varieties. Phytoparasitica 45:461–469

Genç H (2016) Infestations of olive fruit fly, *Bactrocera oleae* (Rossi) (Diptera: Tephritidae), in different olive cultivars in Çanakkale, Turkey. World Acad Sci Eng Technol Int J Biol Biomol Agric Food Biotechnol Eng (WASET) 10:439–442

Genç H, Nation JL (2008) Maintaining *Bactrocera oleae* (Gmelin.) (Diptera: Tephritidae) colony on its natural host in the laboratory. J Pest Sci 81:167–174

Giunti G, Laudani F, Presti EL, Bacchi M, Palmeri V et al (2022) Contact toxicity and ovideterrent activity of three essential oil-based nano-emulsions against the olive fruit fly *Bactrocera oleae*. Horticulturae 8(3):240

Gonçalves MF, Malheiro R, Casal S, Torres L, Pereira JA (2012) Influence of fruit traits on oviposition preference of the olive fly, *Bactrocera oleae* (Rossi) (Diptera: Tephritidae), on three Portuguese olive varieties (Cobrançosa, Madural and Verdeal Transmontana). Sci Hortic 145:127–135

Gümusay B, Olzibey U, Ertem G, Oktar A (1990) Studies on the susceptibility of some important table and oil olive cultivars of Aegean Region to olive fly (*Dacus oleae* Gmel.) in Turkey. Acta Hort 286:359–361

Haniotakis GE (2005) Olive pest control: present status and prospects. IOBC/WPRS Bull 28(9):1–14

Hladnik MA (2017) Review of plant protection against the olive fly (*Bactrocera oleae* (Rossi, 1790) Gmelin) and molecular methods to monitor the insecticide resistance alleles. Acta Agric Slov 109(1):135–146

Iannotta N, Scalercio S (2012) Olive germplasm—the olive cultivation, table olive and olive oil industry in Italy. In: Muzzalupo I (ed) Susceptibility of cultivars to biotic stresses. InTech, Rijeka, pp 81–106

Iannotta N, Noce ME, Ripa V, Scalercio S, Vizzarri V (2007) Assessment of susceptibility of olive cultivars to the *Bactrocera oleae* (Gmelin, 1790) and *Camarosporium dalmaticum* (Thüm.) Zachos & Tzav.-Klon. attacks in Calabria (Southern Italy). J Environ Sci Health Part B Pestic Food Contam Agric Wastes 42(7):789–793

International Olive Council IOC (2022) Newsletter no 175, July 2022

Katsoyannos BI, Kouloussis NA (2001) Captures of the olive fruit fly *Bactrocera oleae* on spheres of different colours. Entomol Exp Appl 100(2):165–172

Konstantopoulou MA, Mazomenos BE (2005) Evaluation of *Beauveria bassiana* and *B. brongniartii* strains and four wild-type fungal species against adults of *Bactrocera oleae* and *Ceratitis capitata*. BioControl 50(2):293–305

Lacey LA, Frutos R, Kaya HK, Vail P (2001) Insect pathogens as biological control agents: do they have a future? Biol Control 21(3):230–248

Malheiro R, Ortiz A, Casal S, Baptista P, Pereira J et al (2015a) Electrophysiological response of *Bactrocera oleae* (Rossi) (Diptera: Tephritidae) adults to olive leaves essential oils from different cultivars and olive tree volatiles. Ind Crops Prod 77:81–88

Malheiro R, Casal S, Pinheiro L, Baptista P, Pereira JA (2018) Olive cultivar and maturation process on the oviposition preference of *Bactrocera oleae* (Rossi) (Diptera: Tephritidae). Bull Entomol Res 109(1):43–53

Margier JF (2012) Guide des productions oléicoles en agriculture biologique. Association Française Interprofessionnelle de l'Olive. Edition: AFINDOL, 31 pp

Medjkouh L, Costa A, Tamendjari A, Bekdouche F, Bouarroudj K (2018) Susceptibility of eight Algerian olive cultivars to *Bactrocera oleae* infestation—a pomological and nutritional quality perspective. Phytoparasitica 46:595–605

Müller FA, Dias NP, Gottschalk MS, Garcia FRM, Nava DE (2019) Potential distribution of *Bactrocera oleae* and the parasitoids *Fopius arisanus* and *Psyttalia concolor*, aiming at classical biological control. Biol Control 132:144–151

Navrozidis E, Zartaloudis Z, Thomidis T, Karagiannidis N, Roubos K et al (2007) Effect of soil plowing and fertilization on the susceptibility of four olive cultivars to the insect *Bactrocera oleae* and the fungi *Sphaeropsis dalmatica* and *Spilocaea oleagina*. Phytopathology 35(5):429–432

Neuenschwander P, Michelakis S (1978) The infestation of *Dacus oleae* (Gmel.) (Diptera, Tephritidae) at harvest time and its influence on yield and quality of olive oil in Crete. J Appl Entomol 86:420–433

Neuenschwander P, Michelakis S (1979) Determination of the lower thermal thresholds and day degree requirements for eggs and larvae of *Dacus oleae* (Gmel.) (Dipt. Tephritidae) under field conditions in Crete, Greece. Bull Soc Ent Suisse 52:57–74

Neuenschwander P, Michelakis S (1981) Olive fruit drop caused by *Dacus oleae* (Gmel.) (Dipt. Tephritidae). J Appl Entomol 91:193–205

Norrbom AL, Carrol LE, Thompson FC, White IM, Freidberg A (1999) Status of knowledge. In: Thompson FC (ed) Expert system and systematic information database, pp 65–251

Ornek ER, Gündoğdu MA, Seker M (2019, May 2–4) Comparison of pomological characteristics of some local olive genotypes [Conference]. 1st international symposium on biodiversity research, Çanakkale, Turkey

Orsini MM, Daane KM, Sime KR, Nelson EH (2007) Mortality of olive fruit fly pupae in California. Biocontrol Sci Technol 17:797–807

Ouguas Y (2021) Mouche de l'olivier *Bactrocera oleae* Gmelin. Institut National de la Recherche Agronomique, 12 pp

Ovruski SM, Schliserman P (2012) Biological control of tephritid fruit flies in Argentina: historical review, current status, and future trends for developing a parasitoid mass-release program. Insects 3:870–888

Pavlidi N, Gioti A, Wybouw N, Dermauw W, Ben-Yosef M et al (2017) Transcriptomic responses of the olive fruit fly *Bactrocera oleae* and its symbiont *Candidatus Erwinia dacicola* to olive feeding. Sci Rep 7:42633

Pereira JA, Alves MR, Casal S, Oliveira MBPP (2004) Effect of olive fruit fly infestation on the quality of olive oil from cultivars Cobrancosa, Madural and Verdeal Transmontana. Ital J Food Sci 16(3):355–365

Ponti L, Cossu QA, Gutierrez AP (2009) Climate warming effects on the *Olea europaea-Bactrocera oleae* system in Mediterranean islands: Sardinia as an example. Glob Change Biol 15(12):2874–2884

Rice ER (2000) Bionomic of the olive fruit fly Bactrocera (Dacus) *oleae*. Univ Calif Plant Prot 10:1–5

Rizzo R, Caleca V (2006) Resistance to the attack of *Bactrocera oleae* (Gmelin) of some Sicilian olive cultivars. In: Proceedings of olivebioteq 2006, second international seminar "biotechnology and quality of olive tree products around the Mediterranean basin", Mazara del Vallo, Marsala, 5–10 Nov 2006, pp 291–298

Rizzo R, Verde GL, Sinacori M, Maggi F, Cappellacci L et al (2020) Developing green insecticides to manage olive fruit flies? Ingestion toxicity of four essential oils in protein baits on *Bactrocera oleae*. Ind Crops Prod 143:111884

Roessler Y (1989) Insecticidal bait and cover sprays. In: Robinson AS, Hopper G (eds) Fruit flies: their biology, natural enemies, and control. Elsevier, Amsterdam, pp 329–336

Sirjani FO, Lewis EE, Kaya HK (2009) Evaluation of entomopathogenic nematodes against the olive fruit fly, *Bactrocera oleae* (Diptera: Tephritidae). Biol Control 48:274–280

Sparks TC, Nauen R (2015) IRAC: mode of action classification and insecticide resistance management. Pestic Biochem Physiol 121:122–128

Torrini G, Mazza G, Benvenuti C, Roversi PF (2017) Susceptibility of olive fruit fly, *Bactrocera oleae* (Diptera: Tephritidae) pupae to entomopathogenic nematodes. J Plant Prot Res 57(3):318–320

Torrini G, Mazza G, Benvenuti C, Simoncini S, Landi S, Frosinini R et al (2020) Entomopathogenic nematodes as potential biocontrol agents against *Bactrocera oleae* (Diptera: Tephritidae). Biocontrol Sci Technol 30:909–919

Tsitsipis JA (1977) Effect of constant temperatures on the eggs of the olive fruit fly, *Dacus oleae* (Diptera: Tephritidae). Ann Zool Ecol Anim 9:133–139

Vignesh S, Chandrasekaran M (2020) Preference of fruit flies, Bactrocera spp. in relation to size and ripeness of intact and fallen guava fruits. Int J Curr Microbiol Appl Sci 9(6):2764–2773

Vontas J, Hernández-Crespo P, Margaritopoulos JT, Ortego F, Feng H et al (2011) Insecticide resistance in Tephritid flies. Pestic Biochem Physiol 100(3):199–205

Yasin S, Rempoulakis P, Nemny-Lavy E, Levi-Zada A, Tsukada M et al (2014) Assessment of lure and kill and mass-trapping methods against the olive fly, *Bactrocera oleae* (Rossi), in desert-like environments in the Eastern Mediterranean. Crop Prot 57:63–70

Yokoyama VY (2012) Olive fruit fly (Diptera: Tephritidae) in California: longevity, oviposition, and development in canning olives in the laboratory and greenhouse. J Econ Entomol 105:186–195

Yokoyama VY (2015) Olive fruit fly (Diptera: Tephritidae) in California table olives, USA: invasion, distribution, and management implications. J Integr Pest Manag 6(1):14

Yokoyama VY, Rendon PA, Sivinski J (2008) *Psyttalia* cf. *concolor* (Hymenoptera: Braconidae) for biological control of olive fruit fly (Diptera: Tephritidae) in California. Environ Entomol 37(3):764–773

Yousef M, Lozano-Tovar MD, Garrido-Jurado I, Quesada-Moraga E (2013) Biocontrol of *Bactrocera oleae* (Diptera: Tephritidae) With *Metarhizium brunneum* and its extracts. J Econ Entomol 106:1118–1125

Zalom FG, Steenwyk RA, Burrack HJ, Johnson MW (2009) Olive fruit fly. UC IPM, University of California. http://ipm.ucanr.edu/PMG/PESTNOTES/pn74112.html

Zygouridis NE, Augustinos AA, Zalom FG, Mathiopoulos KD (2009) Analysis of olive by invasion in California based on microsatellite markers. Heredity 102:402–412

Chapter 5
Climatic Drought in the Hydraulic Basin of Oum Er Rbia, Morocco (1980–2021): Statistical and Cartographic Analysis

Meriem Hamid, Jalal Aguerd, and Mohamed El Ghachi

Abstract Droughts have long been a structural feature of Morocco's climate. Several sequences of drought have hit the country, with adverse environmental and economic consequences. Today, Morocco is experiencing the impacts of climate change, generally manifested in rising temperatures and falling rainfall. Water stress is having a major impact on its major basins, namely the Oum Er Rbia water basin, which is a major reservoir of the country's water resources. The characterization of climatic drought in this basin, through statistical and cartographic analysis, has shown that the evolution of rainfall, during the period 1980–1921, went through five distinct phases, three of which were dry. The severe droughts affecting this basin were recorded during the periods 1983/84, 1992/95, 1998/01, 2019/21, and have marked the entire basin. Examination of the drought years reveals a higher frequency of droughts, with significant spatial extensions. This drought dynamic, particularly that which has marked the last decade, has adversely affected water resources in the Oum Er Rbia basin, and contributed to a water supply crisis in several socio-economic sectors.

Keywords Climatic drought · Global warming · Precipitation and hydraulic basin of Oum Er Rbia

Introduction

Drought refers to a challenge to certain components of the hydrological cycle over a prolonged period. It is one of the extreme climatic phenomena, affecting more people than any other form of natural disaster (Wilhite 2000). Drought can take different forms, namely: hydrological drought, agricultural drought, edaphic drought and climatic drought. The resulting impacts are often harmful and can be disastrous at times (Sebbar et al. 2019).

M. Hamid (✉) · J. Aguerd · M. El Ghachi
Laboratory Landscape Dynamics, Risks and Heritage, Faculty of Letters and Human Sciences, Beni Mellal, Morocco
e-mail: al.ghofrane2011@gmail.com

K. Kahime et al. (eds.), *Climate Change Effects and Sustainability Needs*, Springer Climate, https://doi.org/10.1007/978-3-031-59603-2_5

Drought was always present in the history of Morocco. Several studies have revealed that Morocco has often experienced in the past sequences of severe droughts, which led to famines and epidemics, and marked the Moroccan history by their disastrous impacts (Stour and Agoumi 2009). In the context of climate change, climatic drought has taken on an important dimension, linked to the resulting impacts, particularly in the water resource sector (IPPC 2014, 2022), which is the country's productivity engine. Indeed, Morocco is a Mediterranean country, located between temperate low-pressure systems and the Azores subtropical anticyclone. Its climate is generally semi-arid. Further to its geographical position, the country is threatened by global warming, characterized by a strong recurrence of extreme weather phenomena, especially those related to the decrease in rainfall and the increase in temperature. The recurrence of droughts is then imposed as one of the main climatic hazards threatening the country, and which puts its population at the exposure of numerous risks and challenges, related to the water resource sector, and its contribution in the productivity of the economy. The evaluation of drought, must be done in a permanent way, in order to provide to the decision makers reliable information allowing to evaluate the impacts of the global warming on the country, and to ensure a rational management of the water resource.

Within this framework, the present work aims at characterizing the climatic drought in the hydraulic basin of Oum Er Rbia, during the period (1980–2021). In fact, Oum Er Rbia is one of the large basins in Morocco. It is known for a long time by its richness in water resources, playing a very important role in the development of several socio-economic sectors (Hamid and El Ghachi 2023). The characterization of drought in this basin has already been the subject of several researches, namely: Jouilil et al. (2013) and Daki and al. (2016). But their results do not span the last decade, in which water stress hits the whole basin, and the impacts of global warming have become more accentuated. In addition, their results do not cover the entire basin, due to the limited number of stations studied, which does not correspond to the huge area of the basin.

The present work aims at filling these gaps, based on relatively updated data, going from 1980 to 2021, and covering the whole basin and also its bordering area. It is exactly a study aimed at characterizing the climatic drought in the hydraulic basin of Oum Er Rbia, through a statistical and cartographic analysis, based on the calculation of the index of standardized rainfall and the exploitation of the IDW method of spatial interpolation. The aim of this study is to determine the different sequences of climatic droughts, marking the evolution of rainfall during the period (1980–2021), and to identify their spatial coverage and their environmental and socio-economic impacts.

Materials and Methods

Study Area

The Oum-Er-Rbia water basin is located in the heart of Atlantic Morocco, between 31° and 33° north latitude and between 5° and 8° west longitude. It covers an area of over 48,000 km^2, or 7% of Morocco's total surface area. This hydraulic basin is made up of the OER basin, which slopes to the N.W. towards the Atlantic Ocean, covering an area of 35,000 km^2, and the Atlantic coastal basins to the south-west, covering an area of 13,070 km^2 (Fig. 5.1).

The climate in the basin is arid to semi-arid, with a variety of climatic nuances. It is moderate in the coastal areas, wet in the highlands and arid to semi-arid in the rest of the basin (Hamid and El Ghachi 2023). The hydraulic basin of Oum-Er-Rbia basin is known at the national level by its richness in terms of water resources, whether surface resources or groundwater. This wealth has contributed to the implementation of multiple and diversified hydraulic facilities, playing a very important role in the supply of drinking water for a dozen provinces and also in the supply of water for various economic sectors.

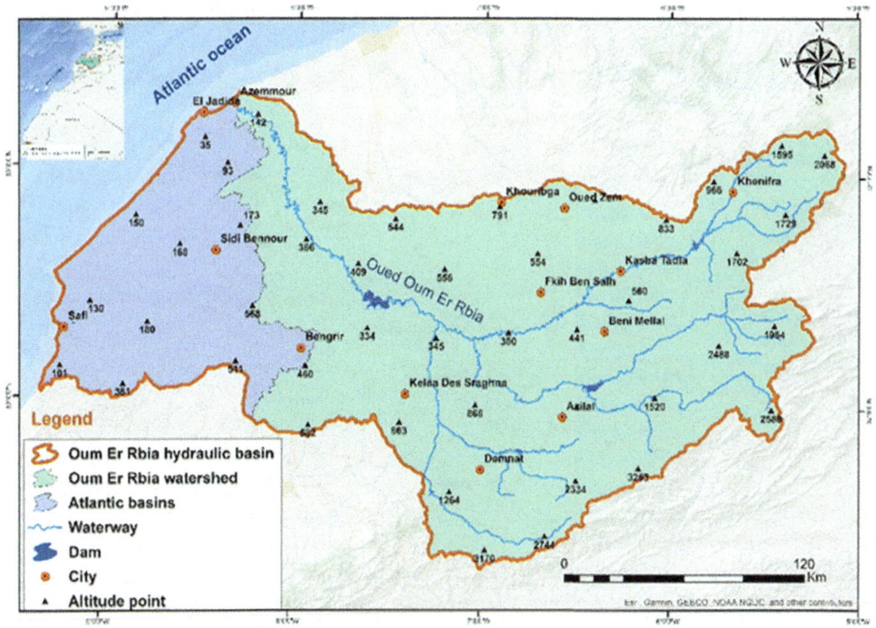

Fig. 5.1 Localization of the hydraulic basin of Oum Er Rbia

Data and Methods

The data base used in this study comes mainly from three sources: The National Meteorological Directorate, the Oum Er Rbia Hydraulic Basin Agency (OERHBA) and the regional offices of agricultural development. It concerns the annual accumulated precipitation of twenty stations, distributed throughout the basin, plus nine adjacent stations, used in the spatialization of results (Table 5.1 and Fig. 5.2). The study covers the period from 1980 to 2021. The processing of this database was carried out by criticizing the data, filling in the gaps and checking the homogeneity, through the application of the Regional Vector Method.

The study of climatic drought can be approached in different ways, depending on the temporal and spatial scale targeted, and on the nature of the available data. In this work, carried out on a medium-term time scale (42 years), and on a basin of a fairly large area, the analysis of climatic drought was done through the standardized precipitation index. In fact, this index was developed by McKee et al (1993). It is used to determine the wet or dry character of a period and to measure meteorological dryness. It also has advantages in terms of statistical consistency (Soro et al. 2014),

Fig. 5.2 Location of the meteorological stations

Table 5.1 Distribution of meteorological gauging stations in the study area

	Station	x	y	Z
Within the basin	Al Jadida	209,186	296,874	28
	Safi	143,333	198,640	81
	Sidi Bennour	222,166	227,298	190
	Al Massira	289,700	211,173	260
	Ouled Sidi Driss	338,948	192,433	334
	Ouaouirnth	344,591	169,896	394
	Mechraa Eddahek	395,036	204,809	425
	Taghzirt	423,134	206,348	573
	Freita	318,284	147,988	575
	Sidi Driss	340,944	140,376	660
	Brg Ahmed EL	454,197	233,217	670
	Bejaad	407,335	241,933	680
	Khouribga	353,046	254,051	770
	My Youssef	323,312	118,414	860
	Taghat	476,192	266,933	880
	Tillouguit	422,684	158,733	1103
	Ait Segmine	361,144	127,288	961
	Tisi Misly	467,872	202,521	1415
	Tamchachat	512,398	274,384	1701
	Ait Tamlil	356,590	95,220	1766
In the area bordering the basin	Ouljat Habboub	420,200	279,300	589
	Sidi Mokhfi	507,584	311,639	1075
	Amougagh	527,593	157,066	1417
	M'Semrir	460,750	111,883	1942
	Tabouazant	541,150	217,450	1640
	Aguelmame S. Ali	537,800	276,000	2078
	Abadla	200,000	129,500	250
	Settat	294,952	291,075	377
	R'DAT	303,100	117,800	690

and also in terms of relevance of the results to short- and medium-term temporal scales.

The equation for calculating the SPI is as follows:

$$SPI = \frac{Xi - Xm}{\sigma}$$

Xi Annual cumulative rainfall

Table 5.2 SPI classification following McKee et al. (1993)

SPI values	Category
≥ 2	Extremely wet
1.5 to 1.99	Severely wet
1 to 1.49	Moderately wet
− 0.99 to 0.99	Near normal
− 1.49 to − 1	Extremely dry
− 1.99.to − 1.5	Severely dry
≤ − 2	Moderately dry

Xm Interannual average rainfall
σ Standard deviation.

The value of the SPI gives a measure of the severity of a wet or dry event as summarized in Table 5.2.

The results obtained are presented cartographically, through the application of the inverse distance weighting (IDW) method. In fact, IDW is a statistical interpolation method, allowing to estimate the value of a variable in a geographical space, from the neighboring observations. This method is based on a deterministic approach and uses mathematical equations to predict the attribute value of a variable at positions where no sample is available, through the linearly weighted combination of a set of sample points.

Results and Discussion

Extraction of Climatic Drought

The first statistical analysis of the SPI results, calculated for each station, shows that during the period 1980–2021, the Oum Er Rbia basin has recorded a remarkable recurrence of dry years (Fig. 5.3). The total number of years with negative SPI values exceeded 21 years at most of the selected stations. This finding indicates that drought is a recurrent climatic phenomenon in the Oum Er Rbia basin. It is thus a structural element of the climate in Morocco. In fact, according to several studies, drought sequences have frequently hit the country for centuries. They have often caused adverse environmental and socio-economic impacts, namely the droughts of 1509, 1604/1605 and 1944/1945.

The results of the IPS, also identified an alternation of dry and wet sequences. The main droughts, which hit the whole basin, were recorded during the periods 1983/84, 1992/95, 1998/01 and 2019/21. These sequences of droughts are also recorded throughout the country. Several studies conducted, either at the level of the whole

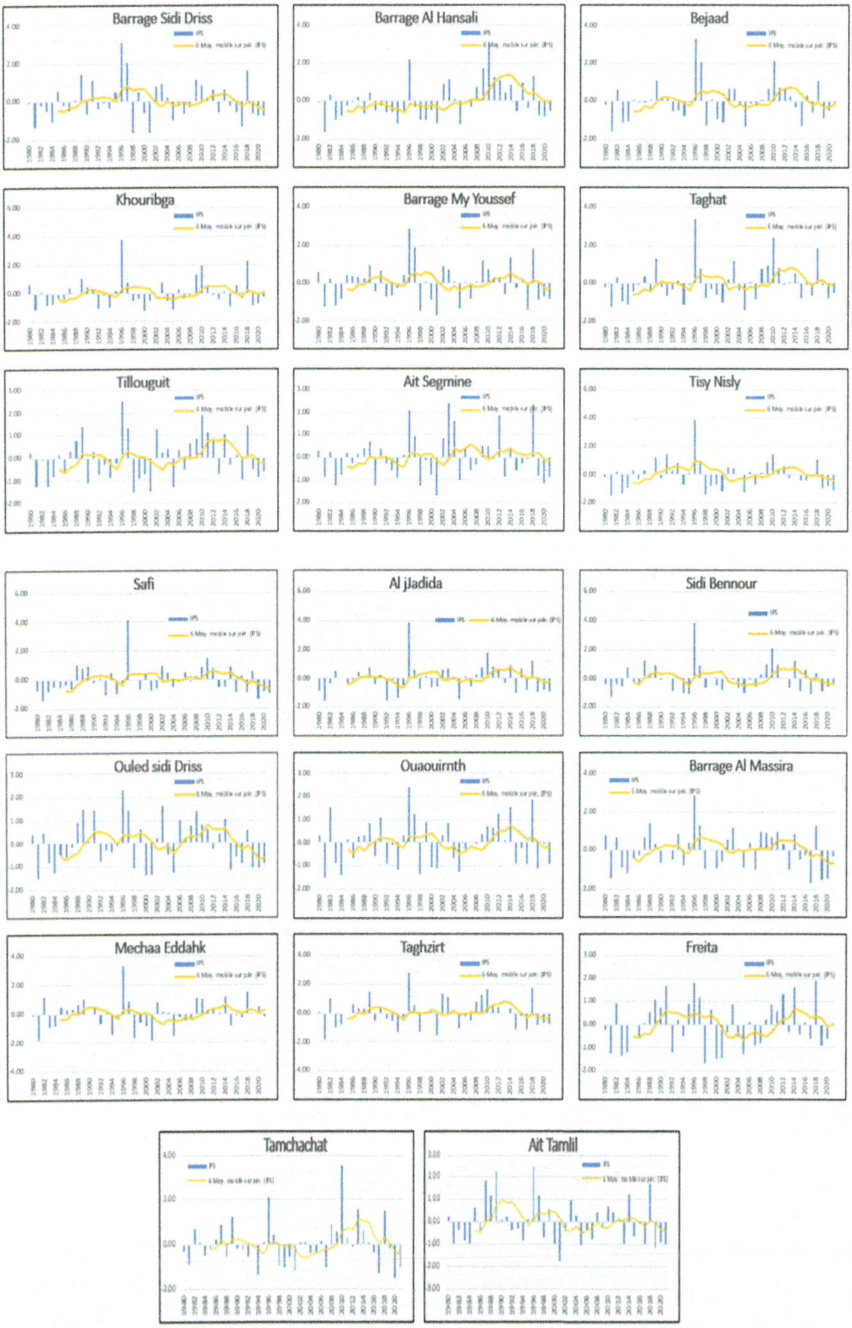

Fig. 5.3 SPI during the period 1980–2021 at the twenty stations located within the OER basin

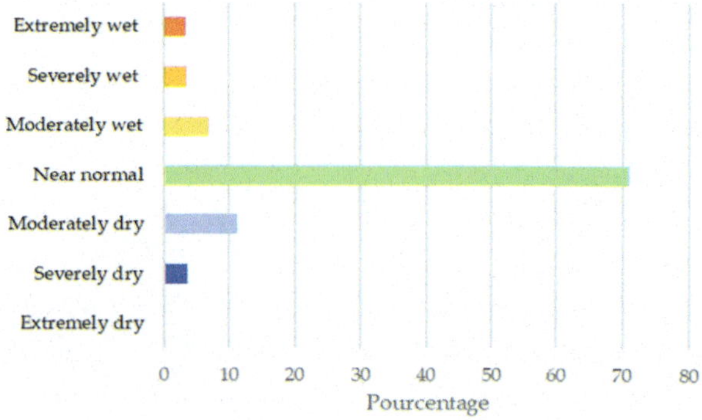

Fig. 5.4 Categories of drought during (1980–2021)

country (Benassi 2008; Sebbar 2013) or on other basins (Mehdaoui et al. 2018; Sebbar et al. 2011), have led to the same results.

Drought Severity

The classification of drought severity according to Mckee's classification, revealed that the majority of dry years recorded, during the period (1980–2021), are generally of moderate drought type (Fig. 5.4). Their number varies between 1 and 8 depending on the station. Their frequency of occurrence is around 1/9. The number of severe drought years reaches more than 4 in several, on the other hand, extremely dry years are not recorded in the period (1980–2021). Severe droughts have mainly marked the years 1981, 1984, 1994, 1998, 2001, 2017 and 2020.

The frequency of drought occurrence in the Oum Er Rbia basin is characterized by spatial disparity. Since the number of dry years varies between 4 and 9. The highest values were recorded at the stations located in the interior of the basin. This spatial variability is due to the diversity of factors influencing the climate in the basin. Namely the maritime influence, the continental influence and the effect of altitude.

Identification and Mapping of Drought Phases

Applying the six-year moving average to the SPI results for each station, the study chronicle was segmented into five fairly distinct phases (Fig. 5.5): two wets, with a few extremely wet years, and three dry phases.

Phase 1 (1980–1987): This is a dry phase, with negative SPI values throughout the basin. This phase is marked by a strong recurrence of dry years, especially during the period (1980–1985);

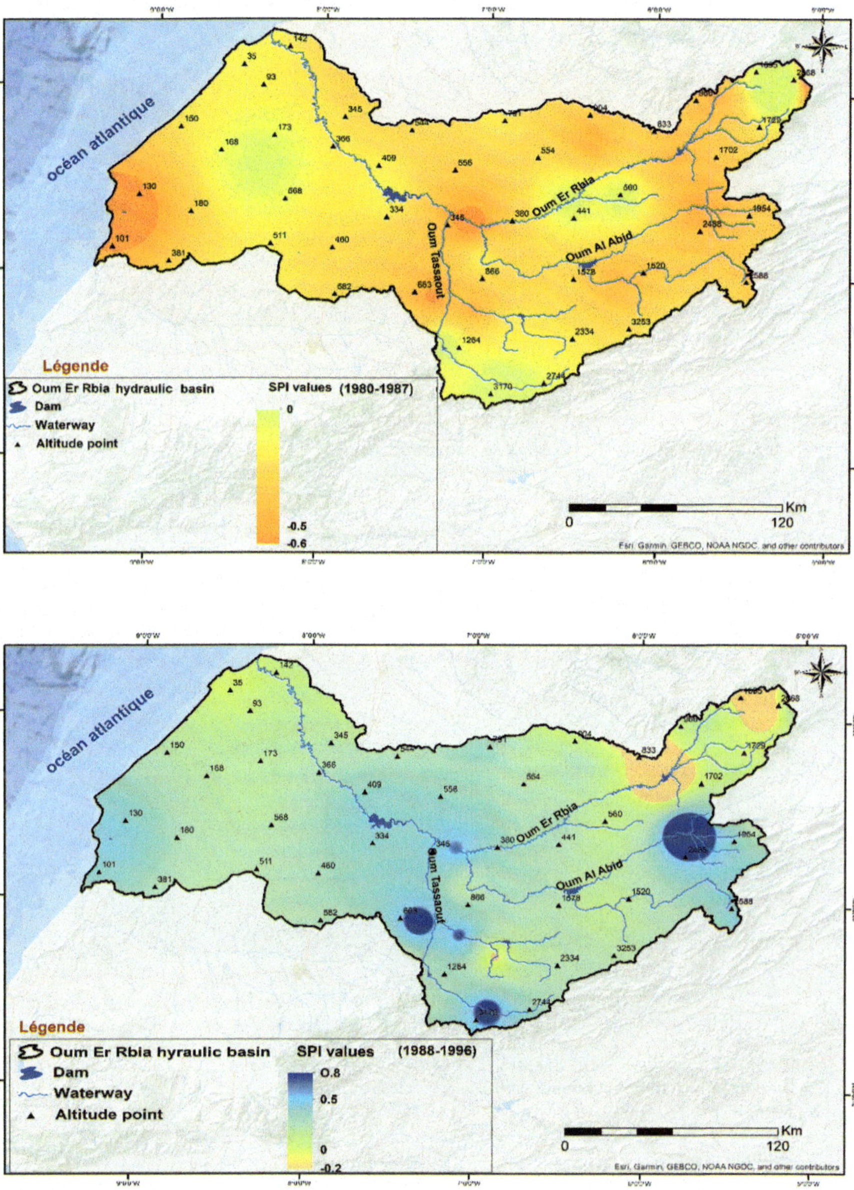

Fig. 5.5 Mapping the results of the SPI during the five phases

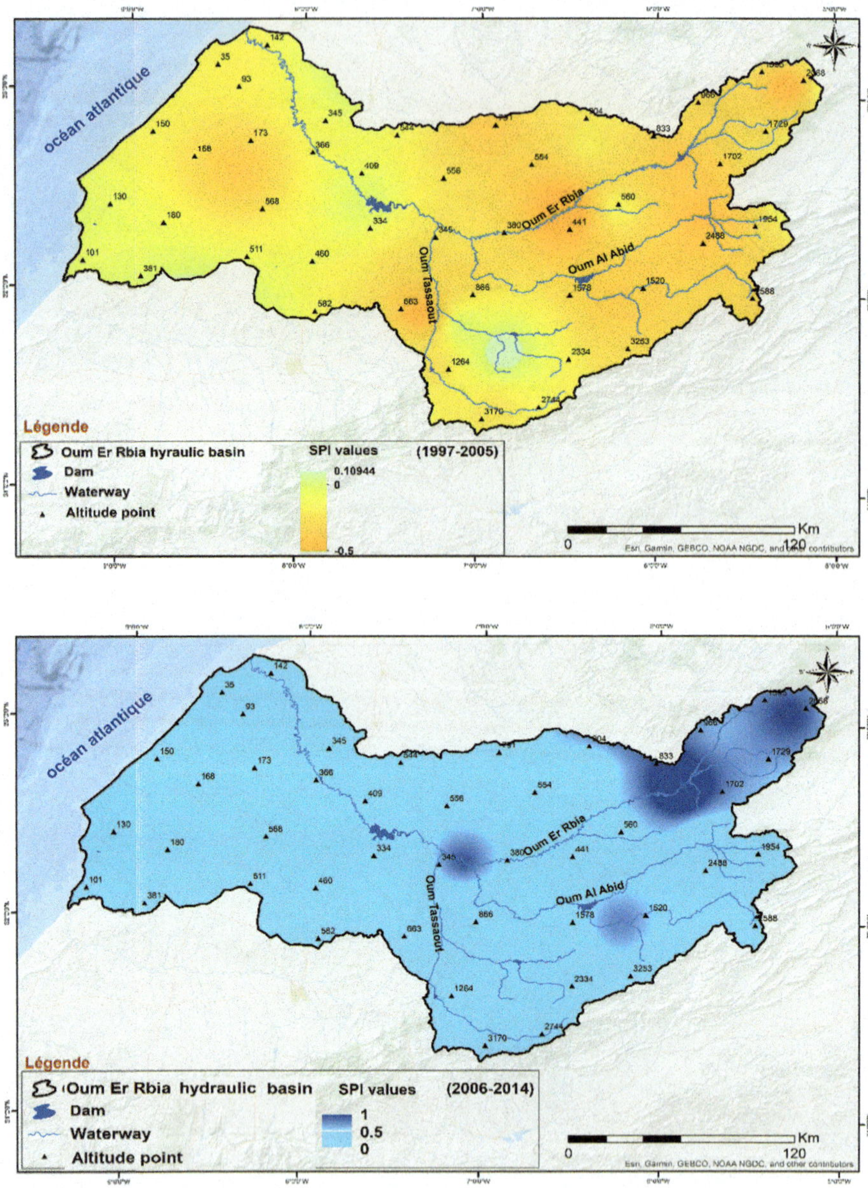

Fig. 5.5 (continued)

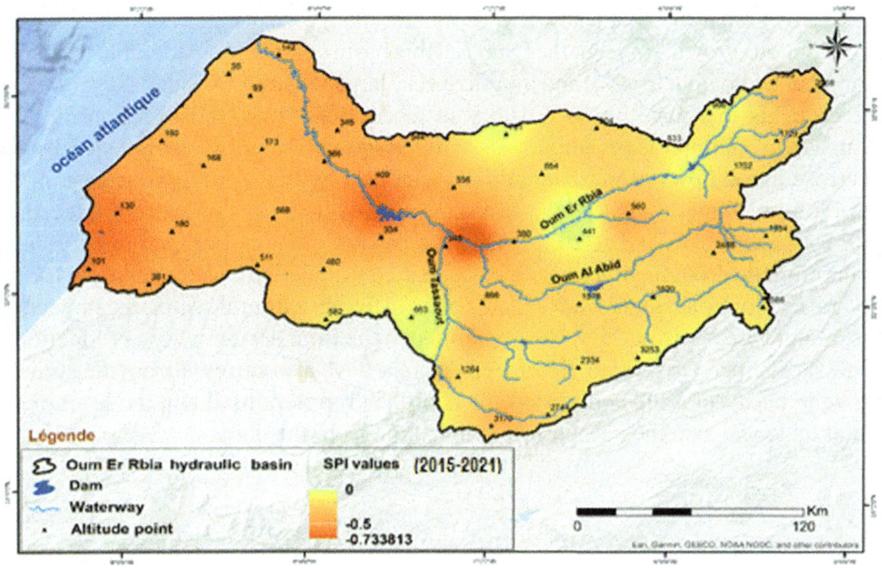

Fig. 5.5 (continued)

Phase 2 (1988–1996): This is a relatively wet phase, extending over a period of 9 years. Its SPI reached 0.32. During this phase, the basin also experienced dry years, namely: 1992 and 1994/95, but their number remains low compared to the wet years;

Phase 3 (1997–2005): During this phase, the SPI recorded a negative value of − 0.22, mainly related to the droughts that hit the whole basin in 1998, 2000/2001 and 2004/2005;

Phase 4 (2006–2014): The wet character marked the climate in the Oum Er Rbia basin during this phase. The SPI indicated positive values throughout the basin, with higher values corresponding to extremely wet years, especially in the upstream part of the basin;

Phase 5 (2015–2021): This is a dry phase. The drought has hit the entire basin hard. The SPI reached − 0.42. This phase is characterized by a remarkable succession of dry years, persisting until now.

Impacts of Drought in the OER Basin

In the basins, whose climate is arid to semi-arid, the persistence of drought for some years, negatively affects their natural resources. In the OER basin, drought is automatically an equivalent to the water challenge, especially in the current situation marked by global warming and socio-economic changes, causing a demand for water in continuous growth. Indeed, the Oum Er Rbia basin contains considerable hydraulic equipment, allowing to store and mobilize very large volumes of water. Nevertheless, the droughts that have hit its territory in recent years have led to a considerable decrease in rainfall. Their impact on water resources has been manifested by the decrease in the flows of permanent rivers and also by the worrying decrease in the reservoirs of dams. In 2022, the filling rate of dam reservoirs did not exceed an average of 15% (OERHBA 2022). This decrease in water resources, largely related to the climatic drought affecting the basin, has contributed to a water supply crisis for several economic sectors located in the basin. In the agricultural sector, the impacts of the drought are harmful. They are not limited to the Bour sector, whose production is necessarily linked to rainfall. The irrigated areas have also suffered from the climatic drought, because of the combined effects of the drop in rainfall and the level of the water table and also the suspension of irrigation from the dam.

Conclusion

The characterization of the climatic drought in the Oum Er Rbia basin, through statistical analysis, based on the calculation of the standardized rainfall index and cartographic analysis, using the IDW method of spatial interpolation, revealed that drought is a frequent climatic phenomenon in this basin. The droughts, affecting the Oum Er Rbia basin, are generally of moderate type. The severe droughts represent only 24% of the droughts recorded during the period 1980–2021. The results obtained have also made it possible to locate three drought sequences. The last one persists until today. The resulting impacts negatively affect the water resource and the economic activities of the population, especially agriculture. The dynamics of drought recorded in the Oum Er Rbia basin and the resulting impacts show that the impacts of the envisaged decrease in precipitation, in the context of global warming, according to the majority of the scenarios considered, would be negative on the Oum Er Rbia basin and also on the whole country. The frequency and severity of drought would increase, and the water supply crisis would worsen. Morocco must focus its efforts, provided within the framework of adaptation to global warming, on the component of drought and water resources.

References

Benassi M (2008) Drought and climate change in Morocco. Analysis of precipitation field and water supply. Options Méditérr 80:83–87

Daki Y, Zahour G, Lachgar R, El Hadi H (2016) Caractérisation de la sècheresse climatique du bassin versant d'Oum Er Rbia (Maroc) par le biais de l'indice de précipitation standardise (SPI). Eur Sci J 12(14):198–209

Hamid M, EL Ghachi M (2023) Variabilité de la température dans le bassin hydraulique d'Oum Er Rbia (Maroc), durant la période (1990–2020): Étude des anomalies et des tendances. Revue Internationale de la Recherche Scientifique (Revue-IRS) 1(2):137–146
IPCC (2014) Climate change 2014: mitigation of climate change. http://www.ipcc.ch
IPCC (2022) Climate change 2022: impacts, adaptation and vulnerability, the working group II contribution. http://www.ipcc.ch
Jouilil I, Bitar K, Salama H, Amraoui A, Mokssit A, Tahiri M (2013) Sécheresse météorologique au bassin hydraulique OUM ER RBIA durant les dernières décennies. LARHYSS Journal P-ISSN 1112-3680/E-ISSN 2521-9782 (12)
McKee TB, Doesken NJ, Kleist J (1993) The relationship of drought frequency and duration to time scales. In: Proceedings of the 8th conference on applied climatology, vol 17, no 22, pp 179–183
Mehdaoui R, Mili EM, Seghir A (2018) Caractérisation à l'aide du SPI de la Sécheresse climatique dans le bassin versant de Ziz (Sud-Est, Maroc). Eur Sci J 14(21):177
Oum Er Rbia hydraulic basin agency (2022). https://www.abhoer.ma
Sebbar A, Badri W, Fougrach H, Hsaine M, Saloui A (2011) Étude de la variabilité du régime pluviométrique au Maroc septentrional (1935–2004). Science Et Changements Planétaires/ Sécheresse 22(3):139–148
Sebbar et al (2019) Caractérisation de la sécheresse climatique au Maroc, Colloque UNESCO-SHF: Sècheresse, Etiages Et Déficits En Eau-Paris, pp 39–47
Sebbar A (2013) Etude de la variabilité et de l'évolution de la pluviométrie au Maroc (1935–2005): Réactualisation de la carte des précipitations. Doctoral thesis, University Hassan II, Faculty. Sci. Ben M'sik, Casablanca, 187 p
Soro GE, Anouman DGL, BI, T. G., Srohorou B, Savane I (2014) Caractérisation des séquences de sècheresse météorologique à diverses échelles de temps en climat de type soudanais: cas de l'extrême nord-ouest de la Cote d'Ivoire. LARHYSS Journal P-ISSN 1112-3680/E-ISSN 2521-9782 (18)
Stour L, Agoumi A (2009) Sécheresse climatique au Maroc durant les dernières décennies. Hydroécol Appl 16:215–232
Wilhite DA (2000) Drought as a natural hazard: concepts and definitions. In: Drought: a global assessment, vol I, chap 1, pp 3–18. https://digitalcommons.unl.edu

Chapter 6
Analysis of Hydrological Regime Variability from 1980 to 2014 of the Source of Ain Asserdoune in Beni Mellal, Morocco

Khouldia Nafia, Mimoun Goumih, and Mohamed El Ghachi

Abstract Water and climate change are regularly cited among the most crises that humanity will have to face in the coming decades. In fact, the links between the two issues are so close that we should perhaps see them as a single issue. In Morocco, in this context of climate change, water resources will be threatened and affected in many regions, with impacts on all areas of development (access to water, food security, health, etc.) while risks of water-related disasters will increase. Those who feel the impacts of these changes the most will be the most vulnerable first. On a regional scale, the Beni Mellal Atlas is known for its high groundwater potential and high unit productivity. Considered as the castle of Morocco, it contains a number of springs that exceed the 54 gauged by the ABHOER (Agence du Bassin Hydraulique de l'Oued Oum Er-Rbia). The source of Ain Asserdoune is a tourist site located in the center of Morocco, at the regional level it belongs to the province of Beni Mellal and at the local level it covers the urban area and part of the municipality of Sidi Jaber. Its module is of the order of 1 m^3/s, the maximum reaches 1.20 m^3/s. The minimum is around 0.76 m^3/s (1980–2014). The flows of this source are collected for the drinking water supply of the urban populations of the city of Beni Mellal, as well as the irrigation of all the fields which occupy the development basin of the source. The main objective of this scientific contribution is to relate the impact of these climate changes on the flows of the source of Ain Asserdoune. To precisely define the characteristics of the flows of the source of Ain Asserdoune, we calculate the CMQ (Monthly flow coefficient) of all the chronicle studied (1980–2017) in order to extract all the information that can inform us about the variability of the hydrological regimes. Thus, we have made an analysis of all the situations of the hydrological regimes of the source of Ain Asserdoune.

Keywords Source of Ain Asserdoune (Morocco) · Flows · Hydrological regimes · Variability · Climate change

---

K. Nafia (✉) · M. Goumih · M. El Ghachi
Department of Geography, Laboratory "Dynamics of Landscapes, Risks and Heritage", Faculty of Letters and Human Sciences, Beni Mellal, Morocco
e-mail: naf.khouldia03@gmail.com

K. Kahime et al. (eds.), *Climate Change Effects and Sustainability Needs*, Springer Climate, https://doi.org/10.1007/978-3-031-59603-2_6

Introduction

Water is a vital natural resource for the survival of humanity and all living species on earth. As a good of creation, water is for all human beings and their communities. One of the most readily available materials on its surface and in its interior, as it is of the most important elements that a person needs in his daily life. Of all the renewable resources on the planet, fresh water is the most relentless for humanity.

However, it has recently begun to pose serious problems for some regions, particularly countries with arid and semi-arid climates such as Morocco. It deserves special attention, since it is very altered and seriously threatened by human activities. Indeed, population growth accompanied by rapid urbanization causes many disturbances to natural environments. The sources are complex hydraulic environments, dependent on many factors (El Ghachi et al. 2011).

In the water sector, officials know that instability and uncertainty are the norm. What is new is the increase in this level of uncertainty caused by climate change. Water is at the heart of these changes, as the observations and projections of the main scientists gathered around the IPCC (Intergovernmental Panel on Climate Change) demonstrate. The hydrological cycle is affected in all its aspects by climate change: changes in precipitation patterns, runoff, sea level rise, desertification processes, etc. In general, disparities in the distribution of water resources will increase: droughts are expected to be more frequent in already semi-arid and arid regions, while water availability is expected to increase in the northern latitudes of the planet. These changes in the water cycle are not without consequences for societies and ecosystems. Water resources will be threatened and affected in many regions, with impacts on all areas of development (access to water, food security, health, etc.) while the risks of water-related disasters will increase. Those who feel the impacts of these changes the most will be the most vulnerable first (The movement of French NGOs committed to access to water and sanitation for all, 2014).

In this context, the source of Ain Asserdoune by its geographical location is subject to a semi-arid climate. This climatic characteristic makes the flow of water an unstable phenomenon and poorly distributed in both time and space.

First, the hydraulic studies conducted on the source of Ain Asserdoune are surprisingly rare. In addition, from the knowledge of hydrology, our interest will be on how the source of Ain Asserdoune is functioned. It is also a question of knowing the way of the hydrological functioning of this resource and the behavior of the flow of the source, as well as to know quels sont leurs rôles à play accentuating or on the contrary weakening the hydro-climatic effects of the source of Ain Asserdoune.

The hydrological regime of a watercourse summarizes all its hydrological characteristics and its mode of variation. It is defined by the variations in its flow usually represented by the graph of the average monthly flow (calculated over several years and called “inter-monthly” flow or monthly module) (Nafia 2019). The hydrological regime of the source of Ain Asserdoune, is a characteristic of the Mediterranean type influenced by karst and by a set of factors, such as relief (altitude, slope …), lithology (permeable rock, impermeable), and climate (temperature and precipitation). It is also characterized by the irregularity of the annual and monthly flow.

The objective is therefore to study the hydrology of the source of Ain Asserdoune. As a first step, a historical study of the flows of the source was conducted in order to understand the annual hydrological regime. Then, a statistical analysis of annual data made it possible to characterize the regime of the source.

General Characteristics of the Source of Ain Assedoune

Presentation of the City of Beni Mellal

The city of Beni Mellal is located in central Morocco, in the Beni Mellal-Khénifra region, which contains the Tadla plain (3600 km^2) in the Oum Er Rbia basin, between the High Atlas to the south and the phosphate plateau to the north.

The urban area of the city of Beni Mellal is a Moroccan urban area centered on the commune of Beni Mellal. It is the largest urban area in the Beni Mellal Khénifra region, both in area and population. The region of Beni Mellal-Khénifra covers an area of 28,374 km^2 or 3.99% of the national territory.

The city of Beni Mellal is the capital of the province of the same name; It falls under the main economic zone. It is located at the foot of Jbel Tassemit on the main road linking Fez to Marrakech. It is located 205 km northeast of Marrakech and 210 km southeast of Casablanca. Its geographical coordinates are: $32°21'$ North latitude and $6°23'$ West latitude. Its administrative boundaries are as follows: to the north, Ouled Ayiche; to the south, Foum Oudi; to the east, Foum Laancer and to the west, Ouled M'Barek and Sidi Jaber.

From a topographical point of view, the city of Beni Mellal is characterized by two essential and distinct elements: (i) To the south: the mountain that dominates the city, (ii) To the north: the fertile plain with its irrigated land. The urbanization was contained at the foot of the mountain, there is currently a regeneration of a Piedmontese urbanization that clings to the mountains, especially in Ourbia, Aït Tislite and Aïn Al Ghazi (Khouyaha 2015) (Fig. 6.1).

The Components of the Source of Ain Asserdoune

The source of Ain Asserdoune is a tourist site where beautiful nature exhibits its magic that plunges visitors into the paradisiacal lands. As for the name of this source, it is linked to an ancestral story very widespread in the circles of the city of Beni Mellal: a man and a woman who were washing wool on the bank of the Oued Laâbid had lost a large quantity. One day, when a man visited Beni Mellal, he heard about this wool coming from the source of Aïn Asserdoune. Driven by curiosity, he tried to unravel the mystery of this wool from the unknown. Back in his village, he managed to find the cave through which water infiltrates to finally arrive at the source of Aïn

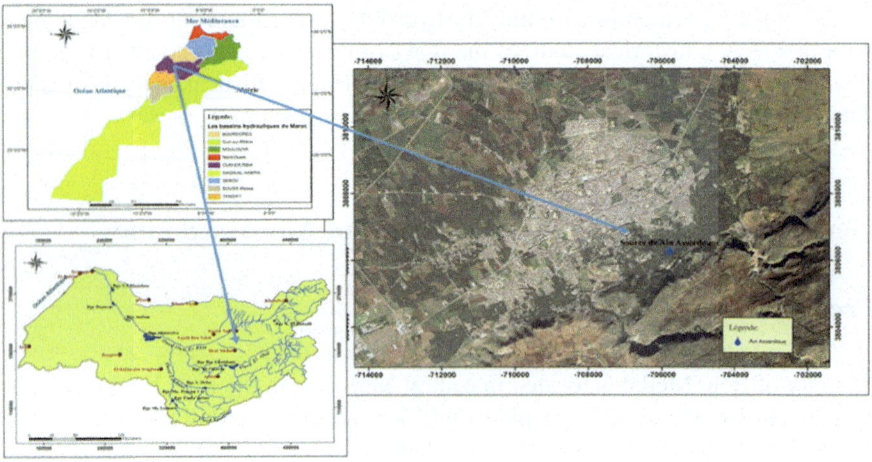

Fig. 6.1 Location of the planning basin at national, regional, and local level. *Source* Nafia (2019)

Asserdoune. He was the only man to make this discovery. One day, he closed the hole through which the water was seeping and noticed that the spring was beginning to dry up. At that time, he decided to negotiate with the personalities of the city the return of water to the source. As water is a source of life, it was decided to grant it a fortune so that the spring would not be dry. The man returned to his village with large sums of money, which he carried on the back of his mule.

The name "Ain Asserdoune" comes from the fortune that this mule had transported, the mule in Berber is "Asserdoune". It was said that after the water returned to the source of Ain Asserdoune, the man who knew the secret of the cave was killed so that the secret would be buried forever.

From a hydrological and hydrogeological point of view, the source of Ain Asserdoune consists of two basins: the catchment area and the development basin or the drainage basin.

The Supply Basin of the Source of Ain Asserdoune

The catchment area refers to the surface area on which water that infiltrates, or runoffs feeds the source. The extension of these surfaces is generally larger than that of the perimeters drained by the source. The catchment area of Ain Asserdoune (Fig. 6.2) covers an area of 222 km and a perimeter of 70 km (Hakim 1982). The Ain Asserdoune catchment area is a karst aquifer where water infiltration and karstification are activated by Atlas tectonics (Tertiary and Quaternary) (Lechhab 1991). It is characterized by the presence of ridges and peaks that reach 2411 m in Ighnayane (Jbel) and 2248 m in Tacemit (Jbel).

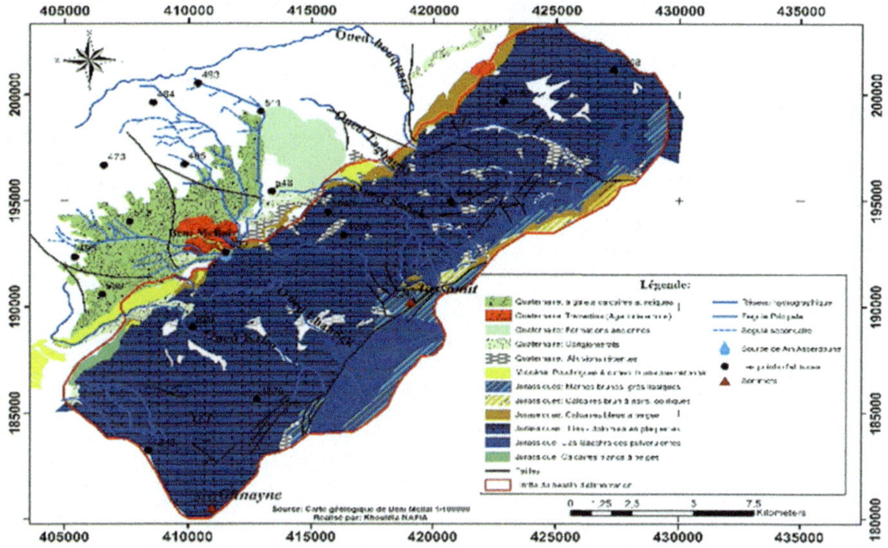

Fig. 6.2 Geology of the Ain Asserdoun catchment area. *Source* Nafia (2019)

The lines of the ridges are generally straight and in a southwest-northeast direction. The karst depressions of great dimension are far from the basin of the source in the regions of Ait Oukabli and Ouaouizaght south of the Atlas of Beni Mellal. This surface karst allows an accumulation and then an infiltration of the surface water that feeds the aquifer of the Ain Asserdoune spring. It is separated from the foothills and the plain by an almost continuous escarpment and cut by the wadis of the Atlas which gave birth to very steep cluses (Lechhab 1991).

The Development Basin of the Source of Ain Asserdoune

The development basin or water discharge basin is most often the drainage site of a spring. In general, these spring basins do not have an outlet and their delimitation depends essentially on the distribution of these spring waters.

The development basin of the source of Ain Asserdoune, covers the urban area of Beni Mellal and part of the municipality of Sidi Jaber. It is characterized by the distribution of waters from the spring of Ain Asserdoune according to traditional customs (Fig. 6.3).

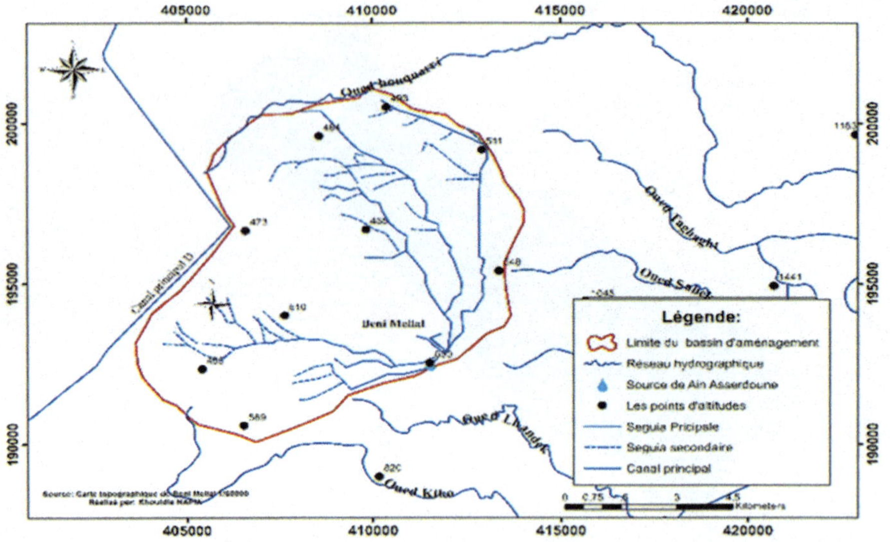

Fig. 6.3 The development basin of the Ain Assedoune spring. *Source* Nafia (2019)

The Climate Framework of the Ain Assedoune Source

Climate is a determining factor in the hydrological cycle, directing the spatio-temporal distribution of water resources and influencing its hydrogeological potential. The geographical position of the source of Ain Asserdoune, gives it a climatic diversity that varies from a humid climate at the summits of the chain of the Atlas of Beni Mellal some peaks of the Middle Atlas, (Basin of the supply of the source) to a Sub-Arid climate in the Dir and the plain below the mountain massifs. Overall, the dominant climate is continental: intense cold in winter with hot summers with high temperature between 43 and 49 (Fig. 6.4).

In addition, rainfall levels have low spatial variability and high temporal variability. There is a succession of well-watered years and dry years. The wettest year is the one recorded in the Taghzirt station, it is the year 1996 (900 mm). The snow appears from 900 m altitude, it covers the summits of the Jbel de Ghnim of Jbel Tasmit. The prevailing wind is the chergui in summer. Average annual temperatures vary between highs of 40 °C and minimums of 2 °C (Data source: the Oum Er Rabia hydraulic basin agency).

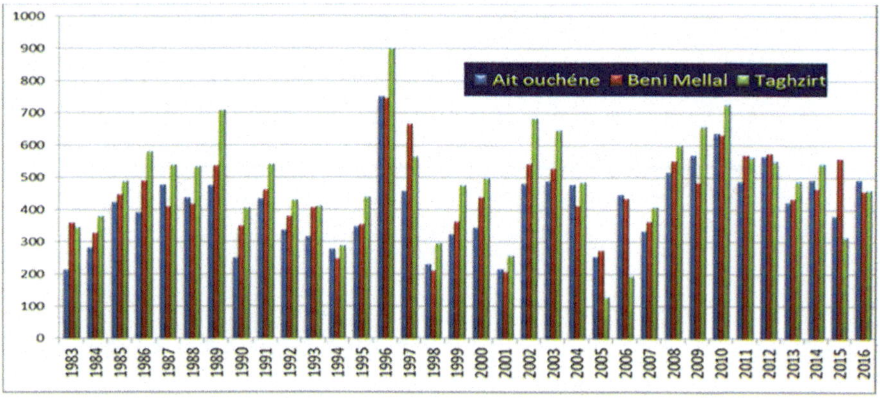

Fig. 6.4 Annual precipitation of Ait Ouchene, Beni Mellal and Taghzirt stations (1983–2016)

Methodology

To analyze the hydrological behavior of the source of Ain Asserdoune, we will study the flows for a 34-year chronicle (1980–2014), in order to extract all the information that can inform us about the modalities of water circulation in the watershed.

To characterize the hydrological regime, hydrological data from the Ain Asserdoune station were also communicated by the Oum Er Rabia hydraulic basin agency, the period chosen extends from 1980 to 2014. To make an analysis of the hydrological regime of the source of Ain Asserdoune.

Results and Discussions

This work first constituted a global approach to monitoring water levels that then allowed us to analyze the flows of the source of Ain Asserdoune to understand the hydrological functioning during the period of low and high water.

The hydrological regime of a watercourse summarizes all of its hydrological characteristics and its mode of variation. It is defined by the average variations of its flow rate as a function of time.

Classification of Hydrological Regime

The classification of hydrological regimes is based on the one hand on the complexity of the distribution of flows during the year (average variations during the year of monthly flow coefficients) and on the other hand on the mode of supply, that is to say the nature and origin of high water (pluvial, snow or glacial).

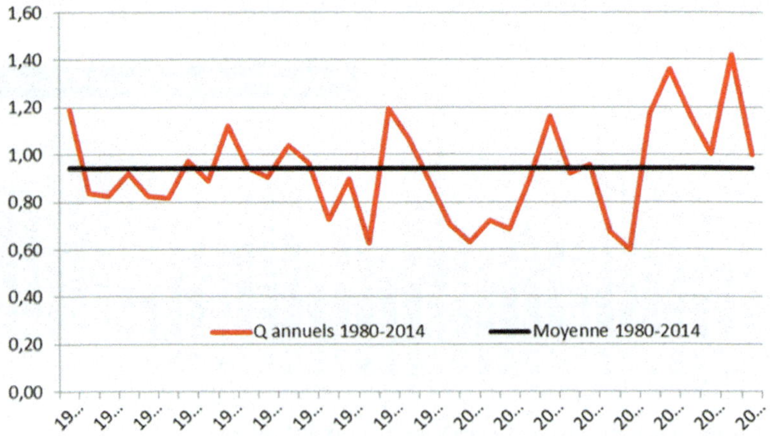

Fig. 6.5 Annual flows relative to the average of the Ain Asserdoune station (1980–2014)

The monthly flow coefficient Cm, is the ratio between the average inter-monthly flow (or monthly module) and the average inter-annual flow (or annual module). It allows to represent the distribution, in percentage, of monthly debits during the year.

$$C_m\ (\%) = \frac{Débit\ mensuel\ moyen}{Module\ Interannuel} \cdot 100$$

The results of the statistical analysis are used to build a scientific confidence rail for the source of Ain Asserdoune.

Annual Variations in Flows from the Ain Asserdoune Source (1980–2014)

It is noted that there is an annual irregularity, over a chronicle of 34 years (1980–2014), the average of the study period is 0.94 m^3/s of which 10 years exceed the average. Similarly, 24 of these 34 years are so-called "deficit" years (Fig. 6.5).

Maximum Annual Flows

At the annual scale, the maximum heights of flows that characterize the so-called chronic (1980–2014) (Fig. 6.6.), show that the maximum flows differ from one year to another and are in the years 1983, 1988, 1996, 2010 and 2013. The maximum of these flow values reached 1.86 m^3/s recorded in 1983. This value is much higher than the modulus.

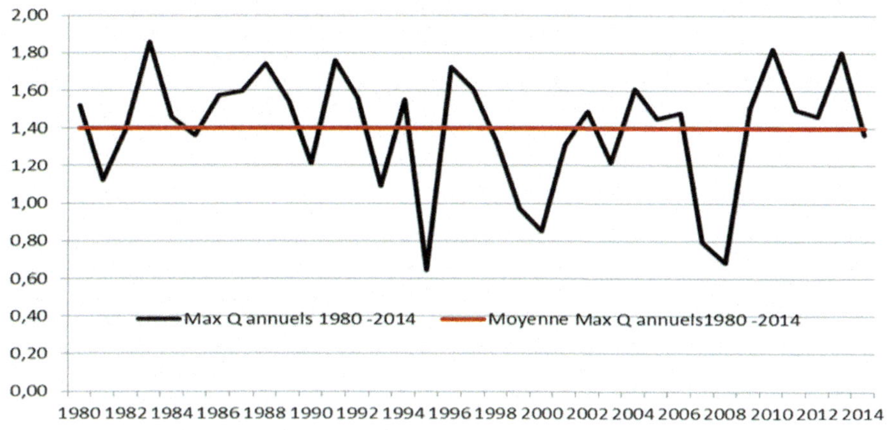

Fig. 6.6 Annual throughput relative to average (1980–2014)

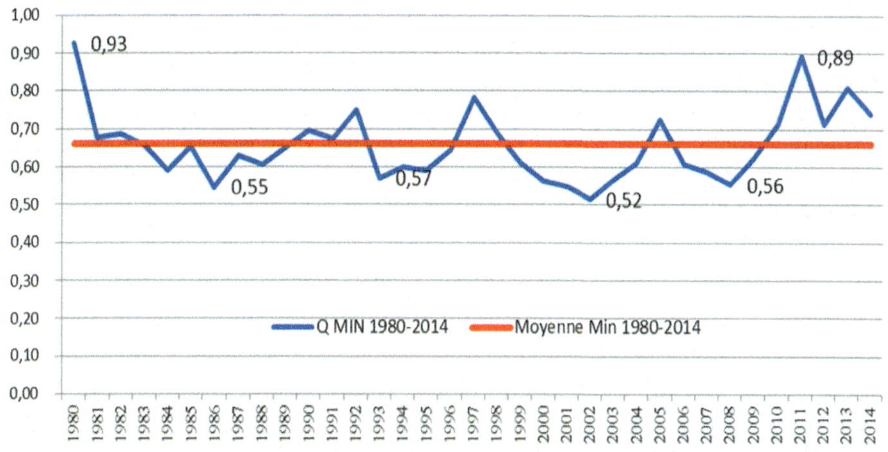

Fig. 6.7 Minimum annual throughput relative to the average (1980–2014)

Minimum Annual Flows

The minimum flow heights (Fig. 6.7) show that the minimum flows differ from year to year and are in the years 1980, 1992, 1997, 2005, 2011 and 2013. The maximum of these flow values reaches 0.52 m^3/s recorded in 2002.

Analysis of Flows from the Ain Asserdoune Source on a Monthly Scale

The monthly and seasonal analysis of the flow of the source of Ain Asserdoune, shows that the average flow knows seasonal variations, even monthly, related to the

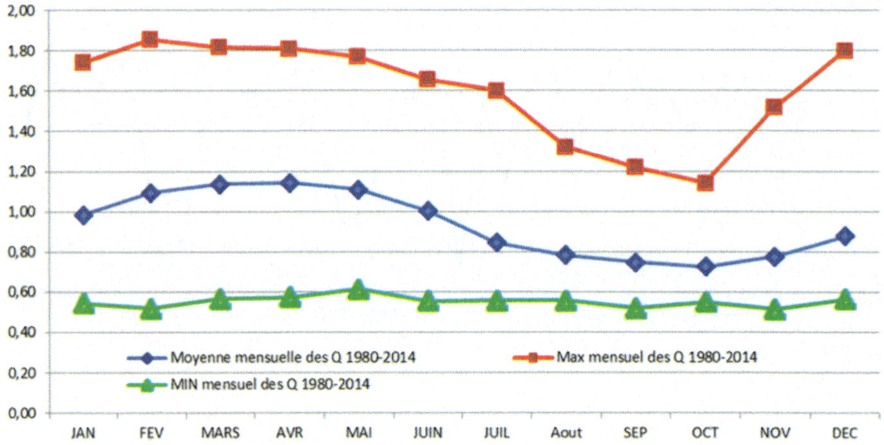

Fig. 6.8 Average monthly flow from the Ain Asserdoune source (1971–2011)

irregularity of rainfall according to the seasons and the duration of rainfall. The figure below shows the average flows for the chronicle (1980–2014) (Fig. 6.8).

The analysis of the graph shows that the maximum average flow was recorded during the month of February (1.86 m^3/s), then this flow begins to decrease until it reaches a minimum value in September (0.51 m^3/s). The average flow remains low between June and October. From this month, it increases to reach an order flow (0.88 m^3/s) in December.

The Hydrological Regime

The regime of the source of Ain Asserdoune is subdivided into two phases:

- The wet phase starts from October until May.
- The dry phase: starts from June to November.

The flow regime of the source of Ain Asserdoune is subdivided into two phases (Fig. 6.9):

The high water phase: begins from January to June. It is characterized by more frequent and intense precipitation, including frequent flow peaks) accompanied by turbidity.

The low water phase: begins from July to December. It is characterized by a more or less regular decrease between April July and October, which corresponds to the approach of seasonal low waters.

The graphs shown in Fig. 6.10 represent all the chronic series years (1980–2014) to distinguish and conclude the periods of low and high water of the source of Ain Asserdoune.

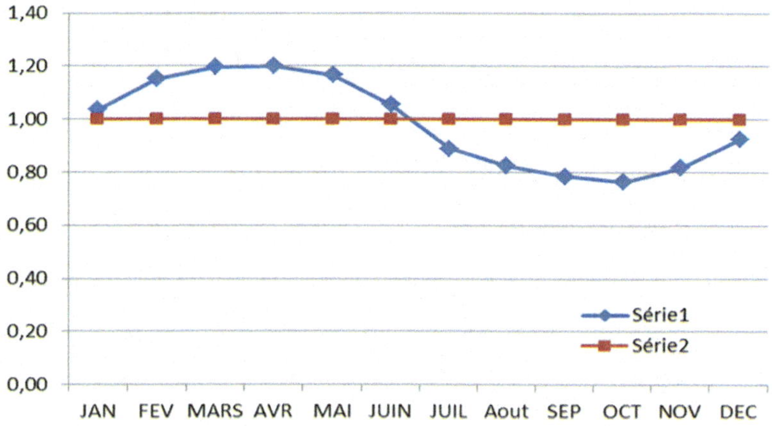

Fig. 6.9 Monthly average module for the basin station (1980–2014)

It is observed that there is an irregularity in the figures, the year is divided into two different parts, the part of high water and that of low water (Fig. 6.11).

The period of high water (January–Jul): during this period the monthly coefficient of flow varies between 1.03 and 1 in February and 1.20; 1.30 in April in this chronic series. During each year of this series, the flow varies slightly as shown in the figure.

The period of high water is characterized by a certain constancy and regularity of flow; the flow increases gradually between February and April and decreases at the same rate between April and July.

The low water period (July–December): during this period, the monthly flow coefficient generally varies below 1 between 0.89; 0.51. The minimum flow is generally recorded in November but low flows can drag on some years until January or February depending on rainfall conditions. During each year of this chronic series the flow varies very slightly.

Spring in general and April in particular are the times of the floods of Ain Asserdoune. In general, the flow of Ain Asserdoune reaches its maximum level and varies slowly during the period of high water.

The period of low water is characterized by a flow that generally tends to decrease. The low water level of autumn generally ends with a significant increase in flow between December and February depending on the year.

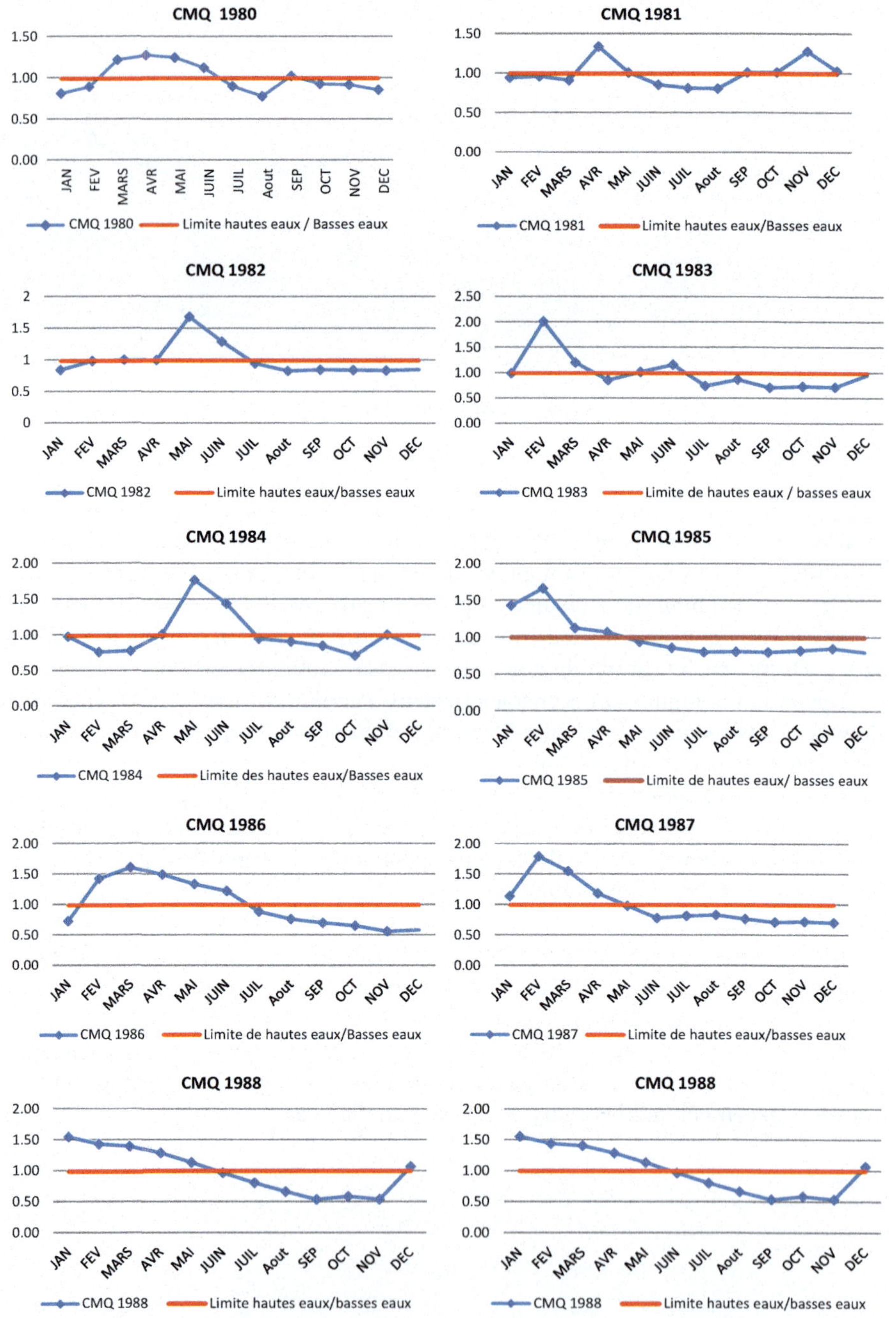

Fig. 6.10 The hydrological regime (1989–2014)

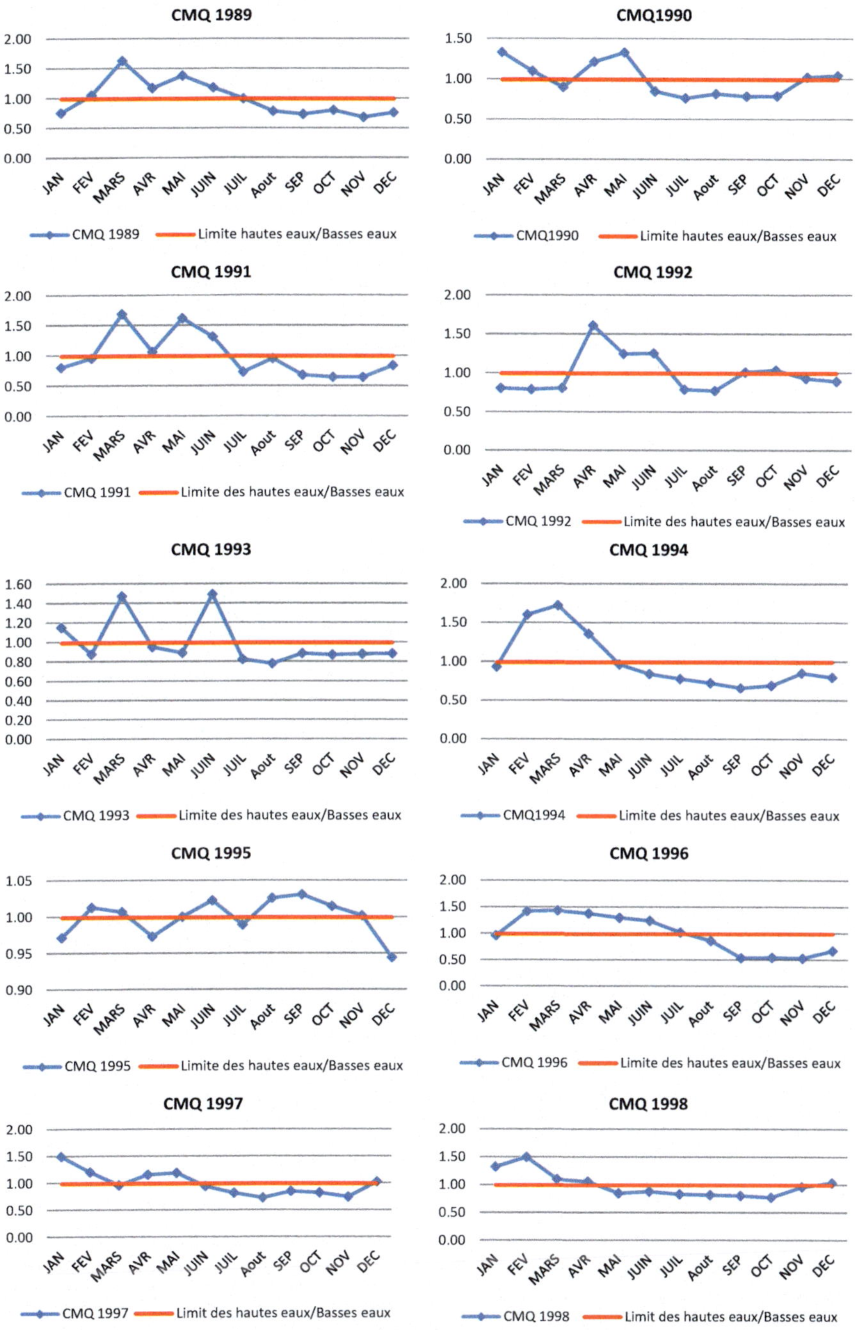

Fig. 6.10 (continued)

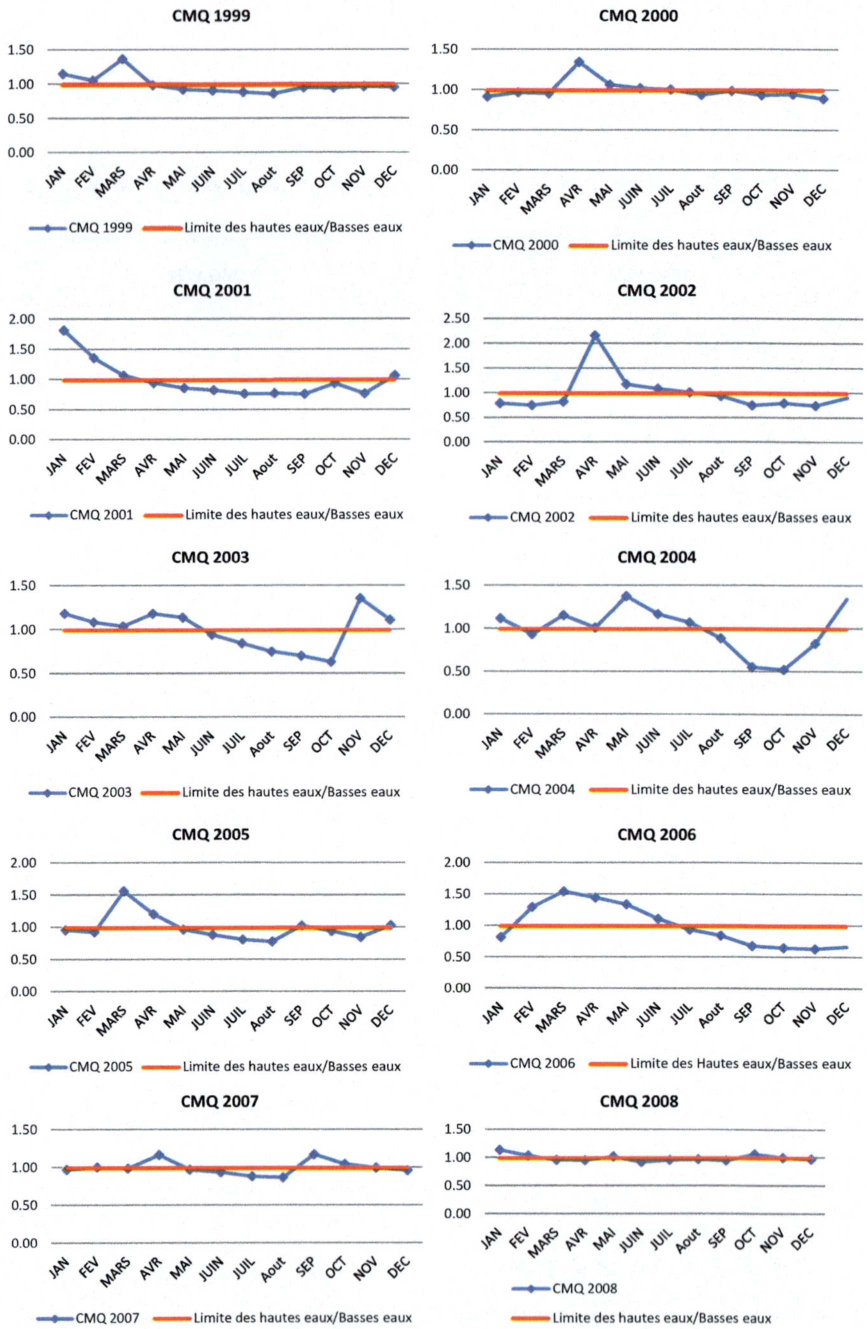

Fig. 6.10 (continued)

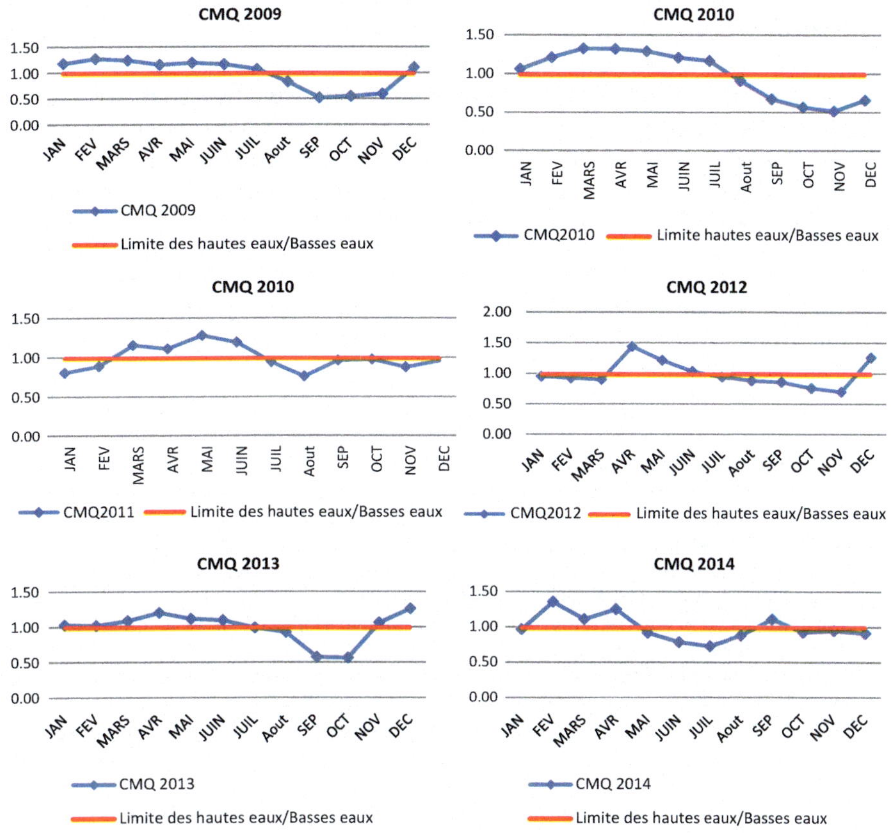

Fig. 6.10 (continued)

Fig. 6.11 Extraction of dry and wet periods

Conclusion

The study of the hydrological framework proposed in this research work differs clearly from the study of chronicles on a weekly scale. From this brief presentation of hydrological data it is shown that the source of Ain Asserdoune characterized by a varied flow due to essentialize les precipitation either liquid or solid. The flow rate of the source increases with wet and dry years.

References

Bienfait P (1978) Note relative aux cavités dans la région du Béni Mellal (Maroc), direction de la région hydraulique du Béni Mellal (D.R.H)

Dubreuil (1974) Initiation à l'Analyse Hydrologique Masson et cie. Editeurs, Paris

El Ghachi M (2007) la seille: un système fluvial anthropisè (Loraine, France) Tom 1. Thése du 3 éme cycle, université Paul verlaine de METZ. France (344 p)

El Ghachi M, Aafir M, Bouberria A (2011) Analyse des tendances pluviométriques dans la ville de Taza, dans un contexte de variabilité climatique. Publication des actes du colloque, 1–115

El Ghachi M, Qadem A, Nafia K (2019) Monitoring and analysis of water levels of Karst springs: case of the spring of Ain Aserdoun (Atlas of Béni-Mellal, Morocco). Int J Environ Water, 1–13. ISSN 2052-3408

El Khalki Y (1990) Etude Hydro-géomorphologique du haut sebou syclinal de skoura et de ses bordures. Thèse du 3éme cycle, Université d'Aix Marseille II. 478 p

Fay G (1999) Du Tafraout au Maider : Techniques et règles coutumières pour l'utilisation des eaux de surface au Maroc, Revue Morsella Tom XXV, n° 3–4, Actes du colloque Franco-Marocain-Eaux et sociétés dans les Montagnes du Maroc et des Pays voisins (p 316)

Guessab (Driss) (1979) Note à propos de l'alimentation en eau potable de la ville de Béni Mellal, (DH de BM)

Hakim B (1982) Recherches Hydrologiques et Hydrochimiques sur quelques karsts méditerranéens. Liban, Syrie et Maroc

Khouyaha R (2015) Les sources dans l'Atlas de Beni Mellal, répartition et exploitation: Cas d'Ain Asserdoune. Mémoire de Master en géographie, Université Sultan Moulay Slimane, Béni Mellal 239 p

Lachhab A (1991) Le rôle de l'eau de la source d'ain asserdoun dans l'organisation de l'espace agricole urbain de Béni Mellal. Thèse de doctorat nouveau régime, université de Nancy 2

Lhoussaine B (1988) Hydrogeologie du Bassin des sources Haut-Atlasien du Dir de Béni Mellal (Maroc)

Mangin A (1975) l'Aquifères karstique sa définition ses caractéristiques et son identification

Marcé A (1975) Contribution des études isotopiques à l'étude des modalités d'Alimentation et de renouvellement des réserves souterraines du Maroc

Martin C (1991) Etat des recherches sur l'hydrologie, l'hydrochimie et la travertinisation actuelle dans le bassin supérieur. Etudes de géographie, physique, Travaux n° XX, URA, CNRS, Aix en provence

Monbaron M (1981) Sedimentation tectonique synsedimentaire et magmatisme basique. L'évolution paleogeographique et structurale de l'Atlas de Béni Mellal. (Maroc)

Nafia K (2019) Les incertitudes et la gestion dynamique des courbes de tarage de la source de Ain Asserdoune: Créations et validations (1978–2018) (Atlas de Beni Mellal—BV de Oum Er Rabia). Mémoire de Master en géographie, Université Sultan Moulay Slimane, Béni Mellal 122 p

Olivier D (1970) Espace Géographie « que-sais-je ? P.U.F n° 1390

Ouhajou L (1996) Espace Hydraulique et société au Maroc, cas des systèmes d'irrigation dans la vallée du Dra. Publications de la Faculté des lettres et sciences humaines, Agadir, Série Thése et Mémoires n° 7

Rollez JP (1973) Sur quelques paleo- reliefs récifaux du Atlas du Béni Mellal. (Maroc)

Taous A (2005) Géomorphodynamique fluvial, processus Morphosédimentaires Paléo-Environnement et mutations récentes des systémes fluviaux. Théses et monographies n° 11, Pub FLSH Sais Fès, Laboratoires d'Analyse Géo-Environnementales et d'Aménagement (425 p)

خويها رشيد (2015) :" العيون بأطلس بني مالل، التو ّزيع واالستغالل: "حالة عين أسردون " . بحث لنيل شهادة الماستر، كلية الاداب والعلوم اإلنسانية، جامعة السلطان موالي سليمان بني مالل.عدد الصحفات 2

Part III
Climate Change and Adaptation: Focus on Agricultural Sector

Chapter 7
The Drought Resilience of Vast Livestock Reared in Agropastoral Versus Sylvo-Agro-Pastoral Systems in Morocco

Bouchra El Amiri, Reda Azmi, Said Laaribya, Brahim Dani, Mounia Sibaoueih, and Berkat Omar

Abstract The present sudy discusses the main climate impacts on rangelands and forest lands and their consequence on the animal feed in two contratsed areas. More precisely, it aims to investigate to what extent the feeding requirements of animals raised extensively are covered in agro-pastoral (Settat region) and sylvo-pastoral systems (Benslimane region)? What are the impacts on livestock under different raifall situations (normal, wet and dry) and what are the most common drought-coping techniques used by farmers in these two systems?. The data collection and surveys were carried out from March 2020 to December 2021. The results showed that for the agro-pastoral system, the forage production was estimated to be 46.8 kg DM, 500 kg DM and 800 kg DM, for wet, normal, and dry years, respectively. In sylvo-pastoral system, the production is higher, ranging between 350 kg DM and 1591 kg DM in both dry and wet years. The field visits showed a decline of rangeland's biodiversity due to the overgrazing combined with the drought impact. In addition, the areas of rangelands and pastures were getting smaller as a result of

B. El Amiri (✉) · M. Sibaoueih
Animal Production Unit, Regional Center Agricultural Research of Settat, National Institute for Agricultural Research (INRA), Avenue Ennasr, P.O. Box 415 Rabat Principal, 10090 Rabat, Morocco
e-mail: bouchra.elamiri@inra.ma

B. El Amiri
African Sustainable Agriculture Research Institute (ASARI), Mohammed VI Polytechnic University (UM6P), 70000 Laâyoune, Morocco

R. Azmi
Center of Urban Systems-CUS, Mohammed VI Polytechnic University of Benguerir, Hay Moulay Rachid Ben Guerir, 43150 Ben Guerir, Morocco

S. Laaribya
Laboratory of Territories, Environment, and Development, Faculty of Humanities and Social Sciences, Ibn Tofail University, 14000 Kenitra, Morocco

B. Dani · B. Omar
Retired Technician of Rangeland Ecology and Management, Hassan II Institute of Agronomy and Veterinary Medicine, Rome, Italy

K. Kahime et al. (eds.), *Climate Change Effects and Sustainability Needs*, Springer Climate, https://doi.org/10.1007/978-3-031-59603-2_7

increased conversion of these areas into cropland and intensive irrigation systems being put in place. From this study, it is to be highlighted that both range and forest lands can no longer meet the needs of the existing flocks.

Keywords Extensive livestock · Climate change · Resilience · Sylvo-pastoral · Agro-pastoral

Introduction

In Morocco, livestock production is central to livelihoods for many rural communities and thus it represents one of the most important pillars of agriculture in the country. It is practiced by 1.1 million farm-households (MAPMDREF 2019) and provides employment to 70% of the rural active population. It also contributes to 36% of total food consumption and satisfies up to 87% of the national demand for dairy products, 98% for red meat, and 100% for poultry meat (MAPMDREF 2019). Most of livestock production is carried out in arid and semi-arid regions of the country practiced under pastoral, agro-pastoral and sylvo-pastoral systems. This suggests the high degree of exposure of livestock to climatic variability. Additionally, as climate-associated risks are becoming more affluent, the vulnerability of the livestock becomes greater. Over the last 50 years, Morocco experienced wide severe successive droughts with significant effects on livestock. To alleviate these negative effects, the government developed several measures and programs geared to livestock protection (Nassif 2008). Besides, farmers also develop their own strategies to cope with successive droughts. In this context, the present study aims to, firstly, investigate to what extent feeding requirements of extensive animals are covered in agro-pastoral and sylvo-pastoral systems within two contrasted climatic areas, namely, Settat and Benslimane while considering different climate scenarios. Secondly, it aims to examine what are the most strategies adopted by the farmers in these two systems to cope with droughts.

Materials and Methods

Overview of the Study Area

The study was carried out in Settat/El Brouj (33° 00′ 00″ N, 7° 37′ 00″ W), known for its agro-pastoral activities, and Benslimane (33° 36′ 44″ N, 7° 07′ 16″ W), known as a sylvo-pastoral system. Both provinces are part of the Casablanca-Settat Region (Fig. 7.1). El Brouj receives 278 mm of rain annually, yet there is significant year-to-year variability, and the average annual temperature varies by season from 18 °C to 34 °C (MADRPM 2008). A substantial amount of variation in annual rainfall

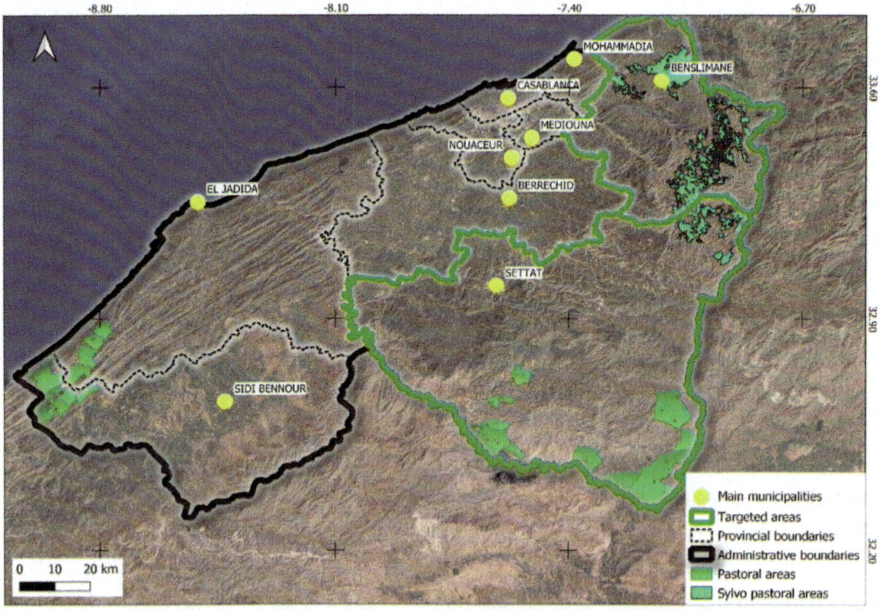

Fig. 7.1 Map showing the location of the study area in the Casablanca-Settat region

was seen by the Benslimane weather station, with a maximum of 724.2 mm and a minimum of 247.9 mm. These areas have a bioclimate that spans from semi-arid to subhumid, with moderate winters that see lows of 7.5 °C and highs of 29.5 °C on average (Bouahi 2010). Figure 7.2 displays the annual rainfall in both regions under three rainfall conditions (Wet, normal, and dry).

Surveys, Data Collection and Analysis

The period of data collection and survey administration took place between March 2020 and December 2021. Figure 7.3 summarizes the main steps for data collection. The initial stages of the procedure involved interviews with relevant specialists and acquiring secondary data from the Agricultural Departments in the relevant areas and local authorities. Afterwords, preliminary fieldwork, in-depth household interviews were done. A 40-person sample of respondents was chosen at random and interviewed using a semi-structured questionnaire. In addition to that, informal discussions with a group of farmers in each of the study areas and with development agents working in those areas were also conducted. Descriptive statistics were used to analyze the collected data. The key topics discussed encompassed areas such as agricultural and livestock yield; the availability of feed resources for livestock production; traditional

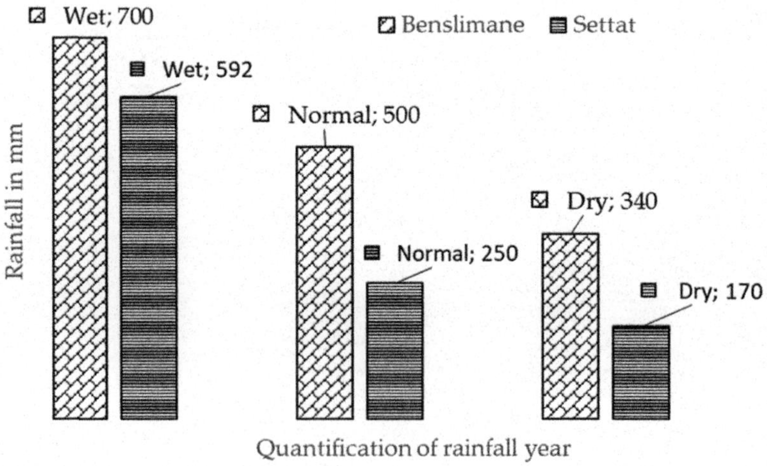

Fig. 7.2 Total annual rainfall in Benslimane and Settat during tree different situations (wet, normal and dry)

livestock feeding and management procedures; and livestock productivity. Strategies employed locally to help livestock cope with drought were also documented.

Forage production of rangelands and forest grazed lands were determined during the period of peak production (March to April 2020, which was characterized as a dry year). Consumable herbaceous standing crop was clipped within 1 m × 1 m plots systematically placed on line transects and using a forced air oven to dried it. Standing crop of the consumable woody plants within reach of the grazing small ruminants was evaluated by determining the average weight of leaves and tender shoots of the standard branch and the density of the concerned shrub species. Forage production during a normal, wet, or a dry year were estimated using rainfall-phytomass models and data from previous field worksrealized in the area.The health of these rangeland ecosystems has been characterized as degraded on a scale of three classes (Good, Vulnerable, Degraded) (National Research Council 1994; Briske 2017).

Dry standing crop was converted to forage unit (FU) considering 1 kg DM (dry matter) = 0.3 FU.

To ascertain the carrying capacity in both sylvo-pastoral (Benslimane) and agro-pastoral (Settat) systems, data related to the total number and variety of livestock were gathered from local authorities. The livestock comprised sheep, goats, and cattle, with the numbers representing average values for the period from 2020 to 2021. To standardize the livestock data, it was deemed helpful to convert the enumerated counts of different species into a singular unit, which is termed a Small Livestock Unit (SLU) (Laaribya et al. 2014). The conversion standard was set as: 1 Sheep equals 1 SLU; 1 Cattle equals 5 SLU; and 1 Goat equals 0.8 SLU.

Several parameters were calculated, such as:

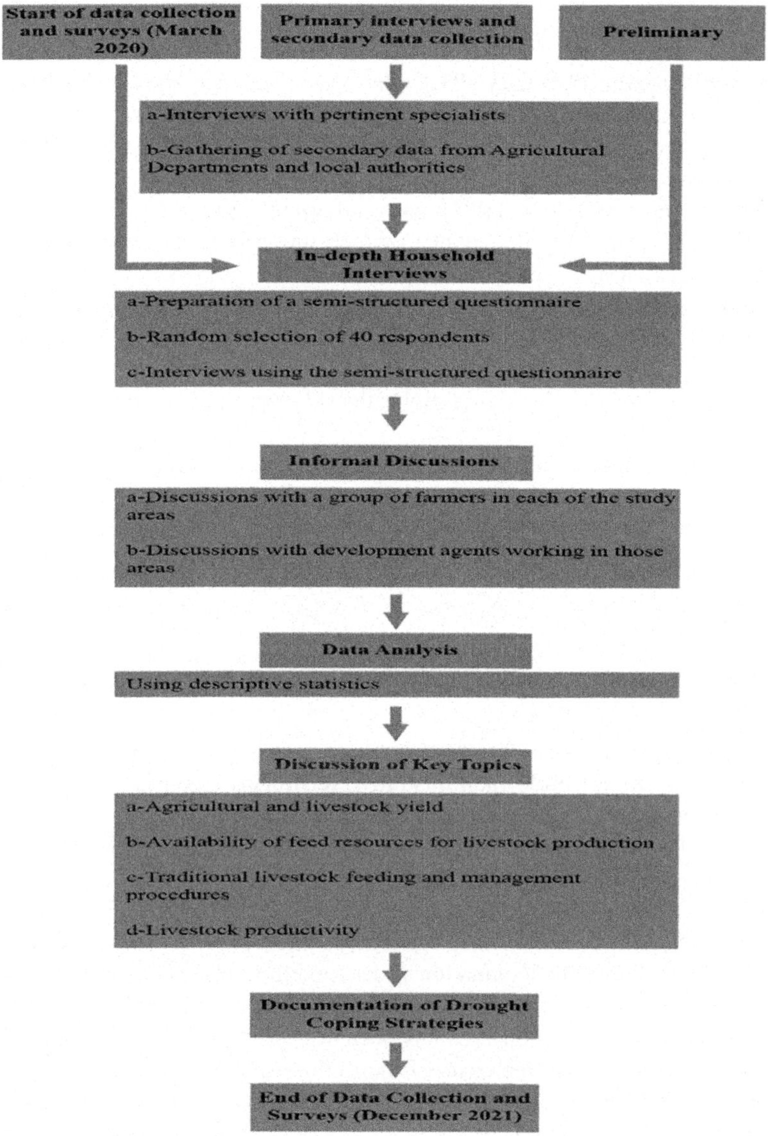

Fig. 7.3 Main steps for data collection to address the goal of the study

Apparent Stocking Rate (ASR) (SLU/Ha): This parameter was calculated by dividing the number of herds (expressed in SLU) by the area (in hectares). It reflects the apparent load of livestock per unit area of land.

Real Stocking Rate (RSR) (SLU/Ha): The Real Grazing Pressure was determined by multiplying the apparent stocking rate by the residence time, then dividing the result by 365 days. This measure provides an indication of the actual grazing load considering the time which livestock spent on the land throughout the year.

Optimal Grazing Capacity (BL): This was calculated by dividing the allowable (for sustained use) Forage Offer (FO) by the Forage Demand (FD). This ratio represents the ideal number of livestock units the land can support without leading to degradation.

Overgrazing Coefficient (O): This was computed as $O = (1 - BL/RSR) * 100$. This coefficient measures the extent to which the grazing capacity of the land is exceeded, contributing to overgrazing.

Forage Deficit (FD): This is the percentage of the overall livestock needs not met by the available forage. It is computed by subtracting the allowable Forage Offer (FO) from the Forage Requirements (FR), dividing the result by the FR, and then multiplying by 100. This deficit highlights the shortfall in forage availability relative to the requirements of the livestock.

In summary, these parameters offer crucial insights into the state of the land, livestock population, and the sustainability of livestock rearing in the study areas. This approach supports the identification of strategies for better land and livestock management to ensure sustainable agricultural practices.

Results

Rangelands within the agro-pasoral system are mostly communal. They are located in the Beni Meskine Cercle (administrative territory) which is the southern area of Settat province and which is characterized by arid climate and by shallow soils (Table 7.1). The northern part of this system (Settat Chaouia plain) has better soil and climatic conditions and is therefore dedicated to crops.

The agro-pastoral system possesses a poor vegetation cover composed mainly of annual species and *Ziziphus lotus*, *Asphodelus microcarpus*, *Asparagus stipularis*. Because of the extent of degradation processes, the palatable and soil protecting perennial grasses such as *Cymbopogon schoenanthus* are rare. The rangeland units, although a proportion of their area has been converted to high-risk annual rainfed crops, were delineated on the basis of their ethnic group users, using the existing official maps and participatory field determination of coordinates with GPS.

Given the relatively homogeneous climatic and edaphic conditions and the homogenizing driving force of degradation, this study found that the 10 communal rangeland units (Fig. 7.4) in Settat area belong to the same rangeland site (*Ziziphus lotus*, *Asphodelus microcarpus*, annuals with a dominance of *Stipa capensis*). The forage production per hectare, determined for the rangeland site, attains 46.8 kg DM, 500 kg DM and 800 kg DM, for a dry, an normal and a wet year respectively (Table 7.2).

Within the National Forest Inventory, ten forest grazed units were distinguished in the Benslimane area (Table 7.1, Fig. 7.5) with a total of 43,457 ha. The largest forest unit has an area of 12,756 ha (Benslimane cork oak forest). These spaces are mostly grazed by a herd composed of 29,743 heads of small ruminants and 3140

Table 7.1 Communal rangeland and units and State forests within Settat versus Benslimane Provinces

Settat		Benslimane	
Rangeland unit	Area in ha	Rangeland unit/Forest	Area in ha
Ain Blal	8096.60	Beni Abid	693
Krakra	18,768.77	Oued Cherrat	3406
Mekrat Jrifa	1698.74	Benslimane	12,756
Mokrat Nkhala Ain Seyada	1678.07	Beninabet-Beniameur	3241
Oukarfa	5059.50	Slamna	4256
Oulad Bouali	9291.60	Ain El Kheil	6477
Oulad Sidi Messaoud	1371.81	Oued Laateuch	2218
Oulad Amer	4798.01	Oued Tifsasine	2615
Oulad Faress	3304.47	Mdakra	10,861
El kalaat	2437.82		
Total	56,505.39	Total	43,457

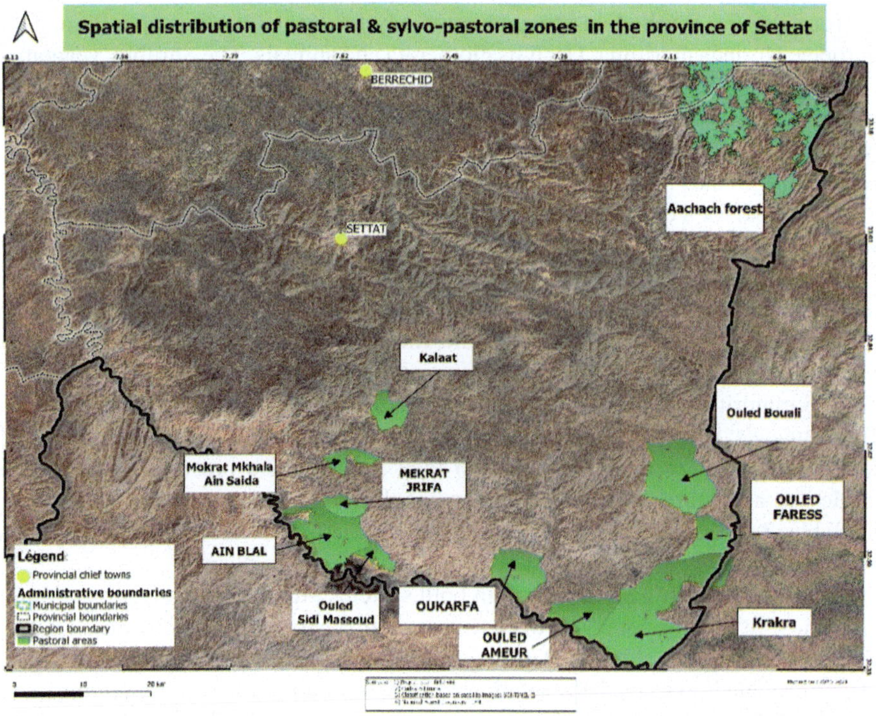

Fig. 7.4 Spatial distribution of pastoral and sylvo-pastoral zones in the Settat area

Table 7.2 The forage production, determined for the rangeland site (Settat province) and for the forest lands

Rangeland site	Area (ha)	Dry year		Normal year		Wet year	
		Standing crop (kg DM/ha)	Total consumable forage (FU)	Standing crop (kg DM/ha)	Total consumable forage (FU)	Standing crop (kg DM/ha)	Total consumable forage (FU)
Ziziphus lotus, Asphodelus microcarpus, annuals	56,505 4	46.8	3,390,324	500	4,237,905	800	6,780,648
Benslimane cork oak forest	23,000	390.1	1,380,000	991	3,450,000	1591	5,520,000
Matorral/ Benslimane	31,859,0	698.2	3,345,195	1110	5,352,312	1410	6,785,967

cattle belonging to 645 breeders. Benslimane cork oak forest includes Quercus suber, Q. rotundifolia, Asphodelus microcarpus, Myrtus sp., Rhus pentaphyllum, Olea europea, Urginea maritima, Daclylis glomerata and annuals. The Shrubby vegetation (Matorral) has an area of 31,859 ha and is composed of Phillyrea angustifolia, Olea europea, Cistus spp. Pistacia lentiscus, Tetraclinis articulata, Rhus pentaphyllum, Myrtus sp, Asphodelus microcarpus and annuals. The forage production, determined for the forest lands attains 390 and 698 kg DM, 991 and 1110 kg DM; and 1591 and 1410 kg DM, for a dry, normal and a wet year respectively in Benslimane cork oask forest and in Matorral (Table 7.2).

Animal feed was frequently purchased from the market, where it could be found among other things in the form of dried beet pulp, cereal bran, and alfalfa hay. Farmers lived in Settat (El Brouj) and Benslimane, the study's two locations, and they kept a variety of animals there, including sheep, cattle, goats, and poultry.

The majority of farmers in El Brouj (Settat) were found to have Sardi sheep. Benslimane's, however, painted a slightly different picture. Not only were the Sardi sheep prominent in this area, but also their crosses. The presence of goats and cattle was also more apparent in Benslimane than in Settat.

Table 7.1 shows how much land is being used, how much forage might be produced, and how many animals are being raised in each place.

A sizable feeding gap was found in both systems after looking at various rainfall scenarios. In Settat as opposed to Benslimane, this gap was more obvious. The gap in Settat occasionally increased during dry years to an alarming 100% as a result of strong drought impact on the vegetation cover composed of annuals. As demonstrated in Table 7.3, the shortfall persisted even in wet years, varying between 50 and 65%.

However, Benslimane's scenario was significantly lessened as a result of the contribution of the woodlands offering a buffer during dry years. As demonstrated in Table 7.3, the feeding gap was significantly minimized, and in some cases, the animals' nutritional requirements were completely satisfied. In certain communes in

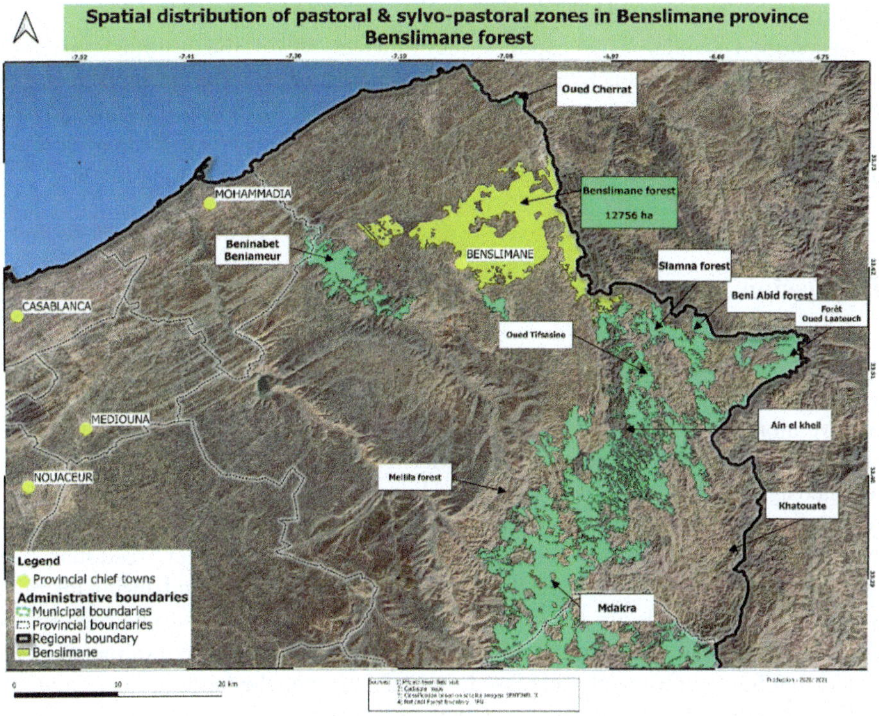

Fig. 7.5 Spatial distribution of pastoral and sylvo-pastoral zones in Benslimane area

Table 7.3 Estimation of livestock feed balance

Province	Lands (ha)	Potential forage (FU/year)	Animal (head)	[a](%)	[b](%)	[c](%)
Settat (agropastoral)	67 907.4	6227 979.75	150 877.6	100	88 to 93	51 to 93
Benslimane (sylvopastoral)	43 457	7 071 159.8	61 876.4	26 to 93	0 to 90	0 to 87

[a]Forage deficit during dry year, [b]Forage deficit during a normal year, [c]Forage deficit during a wet year

Benslimane, the potential for forage even surpassed the dietary requirements of the livestock.

As far as copping strategies are concerned, all farmers believe that they must have a drought coping strategy. The results unveil that farmers are aware of the drought effects and their first reactions is to have feed stocks. As a second reaction, half of the farmers used the social media and interact in forums to know when they can move to look for available pasture. The farmers started adjustment of animal management, purchase of feed from surplus location is a common reaction in both systems. But as soon as the drought becomes severe, the selling of livestock is the most common adaptation and coping strategy. In fact, 75% of the farming households sold at least a part of their livestock. More than 60% of surveyed farmers asked for animal feed (assistance from government/Department of Agriculture supplied fodder or vouchers to buy fodder), while 6% opted for migration. While there is no solution and the drought persists, an emergency slaughter become a common practice.

Discussion

Morocco is characterized by low rainfall and high fluctuation of the precipitation. The country is mainly an arid to semi-arid as more than 93% of its lands did not receive more than 400 mm (Srairi 2017). During the last century, the country experienced an increased frequency of droughts. This climatic phenomenon negatively affected agriculture, forest distribution and livestock production and involved natural resources degradation (Laaribya et al. 2024). The agro-pastoral system seems to be suffering more from this degradation. The forest lands provide more feed than does the range lands only in two areas (Benslimane and Ain El Kheil) and during the rainy years the animal needs are fully satisfied. From this study, it has to be highlighted that both range and forest lands cannot any more meet the feed requirements of the flocks that have significantly increased. In both systems, the lands are experiencing high degradation. In fact, the biodiversity is decreasing rapidly due to the overgrazing caused by very high increase of animal numbers. Besides, rangeland and pastures are shrinking due to the increasing transformation of these areas into cropped lands and the introduction of intensive irrigation systems (Karrou and El Mourid 2008). This statement is not specific to the studied areas as it was also reported in the whole North African region. In fact, Nefzaoui (2002) reported that the contribution of rangelands to livestock diet has decreased from 65 to 10%. The same tendency has been shown in Jordan (Roussa 2002).

Livestock management in the studied area is traditional and extensive, with limited genetic potential (Tsouli Faroukh et al. 2017). Both systems allow poor farmers to increase livestock population and to improve their income. The sylvo-pastoral system appears to be covering the animal requirements more than the other system does. Animals take their feed mainly from the forest and the range lands. During the period of water scarcity (dry years), the stubble and the straw constitute another support for livestock feeding. Other contributions to animal feed resources include fodder crops, fallow land, and supplementary feeding mainly of barley grains, cereal bran, dry beet pulp and oats. Moreover, once the grazing resources have been decreased, farmers are compelled to use delimbing and peeling cork oak trees from the forest lands to support the survival of their livestock (Tsouli Faroukh et al. 2017). Compared to livestock requirements, the agro-pastoral and forest systems cover only a part of the annual needs of grazing livestock. This deficit is partially offset by purchases of livestock feed and excessive harvesting of forest resources, thus contributing to their degradation. Supplementation is offered to livestock during winter and fall, which represent periods of increased livestock needs following the birth and growth of young animals.

The strategies adopted by farmers to cope with droughts, which are reported in this study, are in agreements with the results of Bahta (2020). Besides, Clements et al. (2011) highlighted that farmers could get money whenever they wanted by selling their animals during droughts.

Conclusions

From this study, it has to be highlighted that both range and forest lands are no longer capable of meeting the needs of the flocks that have significantly increased. Looking to the not-too-distant future, climate change will inevitably lead to a significant change in ruminant livestock production systems which must be urgently addressed. The results can be used to identify high priority areas for sylvo-pastoral areas for restoration and conservation against the expected impact of climate change.

Acknowledgements The authors are grateful to the local authorities, regional and provincial agriculture departments (DRA, DPA, ONCA, ONSSA …), the Regional and Provincial Directorates of Water and Forests and combating Desertification of Casablanca and Benslimane as well as local territorial authorities for placing at our disposal the human and material resources as well as the data necessary for the research.

References

Bahta YT (2020) Smallholder livestock farmers coping and adaptation strategies to a gricultural drought. AIMS Agric Food 5(4):964–982. https://doi.org/10.3934/agrfood.2020.4.964

Bouahi S (2010) Usage et conservation des mares temporaires méditerranéennes: cas des mares temporaires de la région de Benslimane (Maroc occidental). Thèse de doctoratenBiologie de l"Université Hassan II Casablanca et de l"Université de Montpellier II

Briske DD (ed) (2017) Rangeland systems. Springer Series on Environmental Management

Clements R, Haggar J, Quezada A, Torres J (2011) UNEP Risø Centre Roskilde. In: Zhu X (ed) Technologies for climate change adaptation—Agriculture sector

Karrou M, El Mourid M (2008) Drought management and planning strategies in semi-arid and arid agro-pastoral of West Asia and North Africa: a review. Options Méditerranéenes Ser A N 80:179–184

Laaribya S, Alaoui A, Ayan S, Dindaroglu T (2024) Changes in the potential distribution of atlas cedar in morocco in the twenty-!rst century according to the emission scenarios of RCP 4,5 and RCP 8,5. Forestist 74(1):16–25

Laaribya S, Alaoui A, Gmira N, Gmira N (2014) Contribution à l'évaluation de la pression pastorale dans la forêt de la Maamora. Parcoursforestiers et surpâturage. Nature & Technology/Nature &Technologie (10)

MADRPM (2008) Le développement de l'élevage au Maroc. Archieve

MAPMDREF (2019) Présentation de la nouvelle stratégie du secteuragricole. Les acquis du Plan Maroc Vert. Eléments de la déclinaison de Green Generation/Filières lait et viandes rouges

Nassif F (2008) The Gender-Livestock-Climate Change connection: local experiences and lessons learned from Morocco. In: Rowlinson P, Steele M, Nefzaoui A (eds) Proceedings international conference livestock and global climate change 2008, 17–20 May 2008, Hammamet, Tunisia

National Research Council (1994) Rangeland health: new methods to classify, inventory, and monitor rangelands. The National Academies Press, Washington, DC. https://doi.org/10.17226/2212

Nefzaoui A (2002) Rangeland management options and individual and community strategies of agro-pastoralists in Central and Southern Tunisia. In: Ngaido T, McCarthy N, Gregorio MD (eds) International conference on policy and institutional options for the management of rangelands in dry areas. Workshop summary paper, CAPRI working paper N° 23, Tunis

Roussan L (2002) Community and household level impacts of institutional options for managing and improving rangeland in the low rainfall areas of Jordan. In: Ngaido T, McCarthy N, Gregorio MD (eds) International conference on policy and institutional options for the management of rangelands in dry areas. Workshop summary paper, CAPRI working paper N° 23, Tunis

Sraïri MT (2017) New challenges for the Moroccan agricultural sector to cope with local and global changes. In: Environmental, social and economic issues of the 21st century, Morocco, pp 165–188

Tsouli Faroukh A, Tellal R, Qarro M (2017) Analysis of socio-economic mutations in the Benslimane forest Lever for sustainable development (Province of Benslimane, Morocco). JMES 8:4415–4425. https://doi.org/10.26872/jmes.2017.8.12.466

Chapter 8
Moroccan Durum Wheat Breeding to Keep Pace with Climate Change

Abdelali Boussakouran, Yahia Rharrabti, Kholoud Kahime, and Mohamed El Yamani

Abstract Morocco is considered one of the most threatened countries by climate variation. Over the last decades, Arable land in this country showed a vulnerability to climate change, which has led to a water deficit, a rise in land salinity, and therefore, a decrease in crop production. In Morocco, the majority of durum wheat areas are located in the rainfed zones, with 50% in the arid and semi-arid regions. It is cultivated in an area of more than one million hectares ranking third after bread wheat and barley. Consequently, extended and severe droughts occurring in the rainy season can lead to significant impacts on both crop yields and agricultural income. Durum wheat yields in Morocco remained low at about 0.9 tons ha^{-1} until the 1970s. With the introduction of improved genotypes, significant increases in yields were observed. However, these grain yield levels are far below both the global world African average. Research findings indicated that, in optimal conditions, the production of durum wheat in Morocco exhibited a positive correlation with the release year of the variety. However, no notable trend was observed in water-limited environments. Hence, when breeders conduct comparable experiments, it is crucial to carefully consider the environmental conditions.

Keywords Climate change · Morocco · Breeding · Durum wheat · Genetic advance · Drought stress

A. Boussakouran (✉) · Y. Rharrabti · M. El Yamani
Laboratory of Natural Resources and Environment, Polydisciplinary Faculty of Taza, Sidi Mohamed Ben Abdellah University, B.P 1223, Taza-Gare, Taza, Morocco
e-mail: Boussakouran@yahoo.fr

K. Kahime · M. El Yamani
Laboratory of Applied Sciences for the Environment and Sustainable Development, Essaouira School of Technology, Cadi Ayyad University, Km 9, Route d'Agadir, BP. 383, Essaouira Al Jadida, Morocco

K. Kahime et al. (eds.), *Climate Change Effects and Sustainability Needs*, Springer Climate, https://doi.org/10.1007/978-3-031-59603-2_8

Introduction

The world population is expected to reach 9.7 billion by 2050. Thus, food production must be doubled because of this population growth and sustainability problems concerning the shrinking of arable lands and reduced water resources (Daszkiewicz 2022). Achieving increased crop yields necessitates substantial inputs, including both organic and inorganic fertilizers, which, as highlighted by Yuan et al. (2017), contribute to the emission of greenhouse gases, thereby causing notable environmental repercussions. In fact, the primary challenge facing durum wheat production is the need to develop viable cropping systems that prioritize environmental sustainability.

Durum wheat yields have increased as a result of the changes in crop management including better seeding equipment, appropriate nutrient application, and effective control of competition from weeds. In addition, genetic improvement has been the key factor to increase crop yields during the past century (Beres et al. 2020). The release of new durum wheat genotypes has improved agronomic traits, disease resistance, and nutritional benefits, which served to maintain the economic attractiveness of durum wheat for producers. Drought is a common characteristic in many durum wheat-growing areas. Thus, breeders attempt to release stable cultivars that are able to maintain acceptable grain yield under water-limited environments.

The adoption of dwarfing cultivars during the second half of the twentieth century after the green revolution produced the greatest advances in enhancing durum wheat (Chairi et al. 2018). Overall, cereal breeding was remarkably successful during the last fifty years, and the improvement effect on durum wheat grain yield and its physiological attributes has been widely reported (Mackay et al. 2021). Examining historical series composed of genotypes that represent the germplasm released in various periods offers valuable insights into comprehending the agronomic and physiological mechanisms responsible for the observed changes in grain yield over time (Boussakouran et al. 2022). In addition, determining the rate of genetic gain allows breeding programs to be regularly evaluated in order to improve their efficiency.

In Morocco, the durum wheat breeding program was initiated in the early 1920s. From the 1970s onwards, the partnership of the National Agricultural Research Institute (INRA) with ICARDA (International Center for Agricultural Research in the Dry Areas) and CIMMYT (International Maize and Wheat Improvement Center) facilitated access to a diverse collection of semi-dwarf germplasm from international sources. This partnership proved highly successful, resulting in the introduction of a significant array of Moroccan wheat varieties. The present chapter addresses the question of whether there has been any genetic advance for durum wheat released in Morocco during the last decades. Our objective in this chapter is to review i) climate change effects on Moroccan durum wheat; ii) Moroccan durum wheat sector; iii) history of Moroccan durum wheat and breeding efforts; and iv) Moroccan genetic advance and its adapting to drought.

Moroccan Durum Wheat Sector

Durum wheat is one of the major food crops in Morocco, it is cultivated in an area of one million hectares, ranking third after bread wheat and barley. The average grain yield during the last ten years (2008–2017) was estimated to 1.74 tons ha^{-1} (Fig. 8.1). In Morocco, durum wheat is particularly important because it is highly appreciated by Moroccan consumers, mainly for the preparation of couscous and other traditional foods. In fact, durum wheat is utilized mainly as flour (refined or whole grain) for the production of a large variety of leavened and flat breads, and for the manufacture of a wide variety of other baking products.

Nowadays, as the population grows and dietary preferences shift towards durum wheat-based products like bread and couscous, Morocco faces challenges in meeting the rising local demand for durum wheat, hence, the country has experienced an escalation in durum wheat imports over the years (Devkota and Yigezu 2021). Historically, durum wheat yields in Morocco remained low at about 0.9 tons ha^{-1} until the 1970s. The introduction of improved wheat varieties in the 1980s led to significant yield increases (Bishaw et al. 2019), reaching a 10-year average of approximately 1.74 tons per hectare after the new millennium (Fig. 8.1). Despite this progress, these yield levels still fall well below the global world average of 3 tons ha^{-1} and the African average of 2.3 tons ha^{-1}.

Generally, cereal in Morocco is grown in a variety of agroecological zones, surrounded by the Atlantic Ocean and the Mediterranean on the West and North, respectively, and by the Sahara on the southeast, with the Western High Atlas on the South. Bread wheat production is concentrated in the Atlantic plains of Morocco, from semi-arid to sub-humid provinces, whereas durum wheat is mainly grown in the semi-arid southwestern plains (Confalonieri et al. 2013).

The primary limitation to wheat cultivation in Morocco stems from drought, particularly during the emergence and post-emergence phases, as well as the grain-filling period. This drought directly impacts seed development. The extent of water

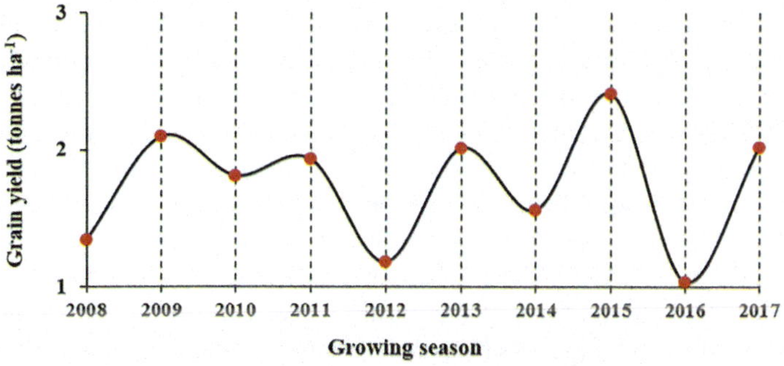

Fig. 8.1 Durum wheat grain yields in Morocco during 2008–2017 period. *Source* Official Data from the Ministry of Agriculture (MAPMDREF 2019)

stress on crop production is significantly determined by both its intensity and duration. If drought occurs in the early stages, grain yield may diminish by up to 25%, while a severe drought affecting the crop from emergence to flowering can result in a staggering reduction of more than 80% (Bregaglio et al. 2015).

History of Moroccan Durum Wheat and Breeding Efforts

The breeder should aim to meet the needs of the entire production chain, which includes farmers, seed handlers, millers, pasta makers, and consumers. To achieve this, collaboration with experts from diverse fields like cereal chemistry, agronomy, pathology, molecular biology, and biometry is essential (Kadkol and Sissons 2016). Wheat breeders strive to develop superior genotypes that exhibit high yields, desirable end-use traits, and resistance to both abiotic and biotic stresses. Historically, public institutions such as government departments of agriculture and universities have been the primary entities engaged in durum wheat breeding. However, private breeding companies have become commonplace in the United States and Europe. The inception of durum wheat breeding dates back to the early decades of the twentieth century in Europe (Xynias et al. 2020).

Italian Nazareno Strampelli played a pivotal role in advancing modern durum wheat breeding (Clarke et al. 2012). In 1923, he released the cultivar "Senatore Capelli," derived from his selection of the local North African landrace "Jean Retifah" in 1915. This cultivar marked a turning point in the cultivation of durum wheat in Italy. Meanwhile, in France, durum wheat research was predominantly conducted in Morocco, Algeria, and Tunisia until the 1960s. The Institut National de la Recherche Agronomique (INRA) in Montpellier and later the Algerian Center for Agronomic, Scientific, and Economic Research (CARASE) coordinated research programs. The introduction of cultivars such as "Oued Zenati 368," "Bidi 17," "Mohamed Ben Bachir," and "Hedba 3" led to a significant increase in durum wheat production. In Morocco, the focus of breeding efforts was on collecting local durum wheat landraces. Subsequent selection programs yielded cultivars like "Zeramek," "Oued Zenati," and "Ky-perounda," released between 1949 and 1956 (Abdelkader 2014).

The foundation of the International Center for Agricultural Research in the Dry Areas (ICARDA) in Syria in 1977 marks a pivotal moment in the advancement of plant breeding. Initially, the center aimed to manage the genetic material and pure lines derived from CIMMYT, specifically focusing on wheat. Subsequently, the ICARDA Improvement Program transitioned to developing crosses and gained autonomy in 2003 (ICARDA 2021). The inaugural success of ICARDA's breeding efforts materialized in the form of the durum wheat cultivar named "Waha," which obtained recognition in Algeria's National Variety Catalogue in 1984. Another noteworthy release was the cultivar "Om Rabi 1," introduced in Morocco in 1989. By 2018, a total of 130 varieties originating from ICARDA's breeding program had been released across 22 countries, showcasing the global impact of its genetic contributions (Bassi and Nachit 2019).

In Morocco, the durum wheat breeding program was commenced in the early 1920s under the Agriculture Research and Development Services. The primary objectives were to enhance both grain yields and quality by leveraging local germplasm introduced from various countries, such as Australia, Algeria, Tunisia, France, India, Italy, Spain, and the USA (Bishaw et al. 2019). Over time, the focus of the breeding program gradually shifted towards hybridization as a means to further improve both yield and quality. Numerous varieties resulting from this program were released and widely cultivated across the country (Jlibene 2005). Moving into the 1950s and 1960s, the key goals of the durum wheat program evolved to were to increase productivity and incorporate disease and pest resistance (specifically rust and the Hessian fly). This was achieved through interspecific hybridization with bread wheat and other related species (Nsarellah et al. 2006).

The establishment of ICARDA in 1977 and CIMMYT in 1966, followed by their collaboration with INRA, facilitated the introduction of enhanced germplasm featuring early maturation and semi-dwarf traits. These partnerships strengthened the breeding program, enabling the release of multiple durum wheat varieties that were not only high-yielding but also resistant to diseases and pests. Since the 1980s, the focus of the durum wheat program in Morocco has shifted towards developing varieties with superior traits such as high yield, excellent grain quality, early maturation, drought and heat tolerance, and resistance to major diseases and pests. Significant efforts and investments have been dedicated to enhancing research capacities in new breeding technologies within the national program over the past decades. These advancements are expected to impact the release of new varieties in the upcoming years. Notably, since 1982, a total of 34 durum wheat varieties, along with other crops, have been officially registered in the national catalog (Fig. 8.2).

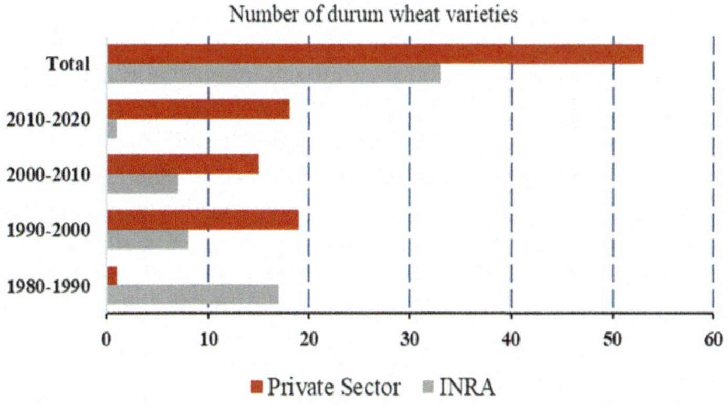

Fig. 8.2 Number of durum wheat varieties registered, 1982–2012. *Source* http://www.onssa.gov.ma

Moroccan Genetic Advance and Its Adapting to Drought

Durum wheat is an important crop in the Mediterranean basin and several large cereal-producing countries in the world. In the Mediterranean countries, the importance of durum wheat is due to good drought and heat tolerance, and its products are deeply connected with the tradition in these countries. About 60% of the world's durum production is concentered in the Mediterranean basin (Abad et al. 2004), it is generally grown under rainfed production systems in the semiarid regions. The crop exhibits a robust yield performance in relatively arid conditions, historically thriving in regions characterized by a dry climate featuring hot days and cool nights throughout the growing season (Royo et al. 2009). The quantity and distribution of rainfall significantly impact both the yield and quality of the crop, and these factors exhibit considerable variability in regions where durum wheat is cultivated. Consequently, drought stress emerges as the primary constraint on crop production. In the Mediterranean region, the cropping of durum wheat traditionally relies on rainfed practices during the wet and cool period from October to April. However, during the subsequent long and dry period from May to September, irrigation becomes essential for viable cropping (Ryan et al. 2008).

In a study conducted in Morocco, Boussakouran et al. (2021) discovered that environmental factors played a significant role in influencing grain yield, yield components, and grain filling traits. The non-irrigated plots exhibited a 30% lower grain yield compared to irrigated ones, with the water deficit causing a 15% reduction in harvest index, a 30% decrease in spikes per m2, and less than a 10% decline in various other traits such as kernels per spike, kernel weight, spikelets per spike, kernels per spikelet, fertile tillering, plant height, and spike length. Similarly, water stress resulted in a roughly 20% decrease in grain filling traits. The observed reductions in grain yield and associated traits during stress conditions were attributed to factors such as pollen sterility, accelerated leaf senescence, oxidative damage to photosynthetic machinery, decreased rates of carbon fixation and assimilate translocation, as well as impaired grain development (Farooq et al. 2014).

In addition, a strong impact of environmental conditions on physiological, biochemical and morphological characteristics has been detected in Moroccan durum wheat developed over three decades among growing seasons, particularly in rain-fed trials (Boussakouran et al. 2022). In fact, all physiological traits declined steadily during the driest growing seasons, as stress became more severe. The shortage of water that takes place during dry growing seasons affects many of the metabolic changes that occur in all plant tissues, including alterations in cell volume, loss of turgor, disruption of membrane integrity, solute concentration and protein denaturation. The impact of year-to-year climatic variations emerged as a key impediment to the formation of grain yield. Clear distinctions were observed in the majority of assessed traits across the three growing seasons. Grain yield exhibited significant variability among these seasons, primarily attributed to differences in total rainfall and its distribution throughout the vegetative and grain-filling periods (Cossani et al. 2009).

Increased grain yield and improvement of other agronomic traits are important objectives for breeders. Plant breeding in cereals, and particularly in durum wheat, was remarkably successful during the second half of the twentieth century (De vita et al. 2010). In fact, the global durum wheat yield has significantly increased due to the improvement of varietal performance. The testing of varieties bred at different periods in a common environment gives a direct estimate of the genetic progress (Giunta et al. 2007). Assessing the rate of genetic improvement in breeding programs stands as a crucial metric for monitoring strides towards the objective of ensuring food security in the forthcoming years (Rife et al. 2019).

Several breeding programs have achieved significant genetic progress in grain yield durum wheat (Chairi et al. 2019; Del Pozo et al., 2019), with an absolute genetic gain ranged between 20 and 72 kgh^{-1} per year. In Morocco, a similar genetic progress for grain yield (23 kg ha^{-1} $year^{-1}$) was observed when the data were employed for linear regression analysis (Boussakouran et al., 2021). Nevertheless, genetic change in grain yield exhibited variations in rainfed and irrigated trials (Fig. 8.3). Specifically, grain yield demonstrated an increase over the years under favorable conditions, whereas no significant trend was observed in water-limited environments. The findings of Boussakouran et al. (2021) underscored that the grain yield of modern varieties showed greater sensitivity to water deficit. This suggests that the estimated genetic change was significantly influenced by environmental conditions, aligning with the conclusions of Mohammadi (2012), who, in a study involving 14 durum wheat genotypes across 21 environments, noted that the most substantial genetic progress occurred in stress-free environments. A similar observation was made by Sanchez-Garcia et al. (2013), who highlighted the dependence of genetic progress in Spanish cultivars on environmental conditions. These overall results emphasize the complex interaction between genetic changes in cereal yield and the specific environmental context studied. Thus, the breeding effects measured for a trait in one agro-ecological condition may not necessarily be replicated in other conditions. Therefore, when conducting similar experiments, breeders should exercise particular caution and account for the diverse environmental conditions that may impact the outcomes.

Climate Change Effects on Moroccan Durum Wheat

In the coming century, climate changes, primarily linked to the rise in greenhouse gases, are expected to induce global warming in specific regions. In Mediterranean environments, cereal crops are grown predominantly in the semiarid and subhumid zones. With climate change, an increase in temperature, in the length of the summer dry period, and in its extension are expected. These changes are expected to intensify the impacts of drought, as very dry soil stimulates the sensible heat flux, which can interact with the atmosphere, intensifying heat waves (Yves et al. 2020). Consequently, vegetation will experience prolonged water stress. Since the 1960s, drought events have stood out as significant contributors to global crop failures (Cottrell

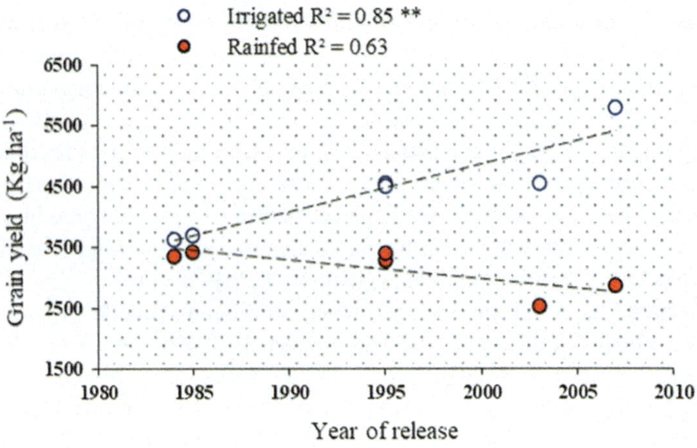

Fig. 8.3 Regression of grain yield on year of release in six Moroccan durum wheat varieties grown under rainfed and irrigated conditions. Each point represents the mean value of one variety during three growing seasons (Boussakouran et al. 2022)

et al. 2019), exerting a substantial influence on the world economy and food security. Morocco, with 90% of its cereal areas situated in rainfed zones and half in arid and semi-arid regions, is especially susceptible. Nationally, durum wheat, ranking third after bread wheat and barley, is cultivated across over a million hectares (Ben-Amar et al. 2020). As a result, extended and severe droughts during the rainy season pose a serious threat to crop production and agricultural revenues in the country.

Soil moisture serves as a pivotal factor within the soil–vegetation–atmosphere system, functioning both as a switch and integrator for inter-connected water fluxes. It plays a crucial role in regulating energy flows and biogeochemical cycles, thereby exerting a significant influence on the process of plant growth (Teuling et al. 2013). The socio-economic impacts related to drought may be severe, as they often build up slowly and over a long period, and can persist for years once the meteorological drought has ended (Vrochidou et al. 2013). Drought stress is one of the most adverse factors that severely reduces crop yield, it decreases CO_2 assimilation rates, leaf sizes, stem extension, water-use efficiency, assimilate partitioning, and activities of sucrose and starch synthesis enzymes (Anjum et al. 2011).

For improving grain yields under water-limited conditions, the development of new durum wheat varieties with high yield potential through identifying drought tolerance mechanisms is of great importance. To facilitate the development of tolerant varieties that can survive and give better yields under drought conditions, a thorough understanding of the various morphological, phenological, and physiological characteristics that govern the yield under water-limited conditions is a prerequisite (Reynolds et al. 2005). Important target characteristics are the decrease in plant size with a heightened harvest index, and the reduction in number of days to anthesis and maturity enabling the crop to escape terminal drought stress (Blum 2010); as well as the architectural characteristics of the roots, such as distribution and root length density, which enable efficient water uptake (Ehdaie et al. 2012). In addition, traits associated with reduced evaporative losses and photo-assimilate production such as leaf rolling, flag leaf persistence, stomatal conductance, and canopy temperature must be selected (Dodig et al. 2012). Moreover, selection based on grain yield should be supported by proper calculation, utilization, and interpretation of various drought indices that evaluate genotypic yield response to water stress.

Conclusion

Durum wheat cultivation in Morocco has immense socio-economic relevance and is a key contributor to the national food security. However, its practice is closely linked to rainfall conditions, which are being increasingly subjected to the effects of climate change, thus increasing the vulnerability of the whole agricultural sector. In light of this, adjustment of agricultural practices, particularly those related to durum wheat growing, as well as the selection and introduction of resilient varieties are proving to be fundamental measures for dealing with worsening extreme climatic events. Such measures are crucial to guarantee the long-term sustainability of durum wheat growing in Morocco faced with the climatic issues of the future. Meanwhile, an innovative and integrated approach, in which scientists, government officials and farmers work together, is needed to build collective capacity to meet these climate challenges.

References

Abad A, Lloveras J, Michelena A (2004) Nitrogen fertilization and foliar urea effects on durum wheat yield and quality and on residual soil nitrate in irrigated Mediterranean conditions. Field Crop Res 87(2–3):257–269

Abdelkader B (2014) The history of wheat breeding in Algeria. In: Porceddu E, Damania AB, Qualset CO, (eds) Série a Mediterranean Seminars, No 110, Options Méditerranéennes, Proceedings of the international symposium on genetics and breeding of Durum Wheat, Rome, Italy, 27–30 May 2013, CIHEAM: Bari, Italy, 2014, pp 363–370

Anjum SA, Xie XY, Wang LC, Saleem MF, Man C, Lei W (2011) Morphological, physiological and biochemical responses of plants to drought stress. Afr J Agric Res 6(9):2026–2032

Bassi FM, Nachit MM (2019) Genetic gain for yield and allelic diversity over 35 years of durum wheat breeding at ICARDA. Crop Breed Genet Genom 1:1–19

Ben-Amar A, Nsarellah NE, Bouizgaren A, Mouradi M, Mahboub S, El Bouhmadi K (2020) Assessment of leaf rolling role in maintaining agrophysiological performances of Moroccan durum wheat cultivars under water deficit conditions. Plant Physiol Rep 25(3):436–443

Beres BL, Rahmani E, Clarke JM, Grassini P, Pozniak CJ, Geddes CM, Ransom JK et al (2020) A systematic review of durum wheat: enhancing production systems by exploring genotype, environment, and management (G× E× M) synergies. Front Plant Sci 11:568657

Blum A (2010) Plant breeding for water-limited environments. Springer Science & Business Media

Boussakouran A, El Yamani M, Sakar EH, Rharrabti Y (2021) Genetic advance and grain yield stability of moroccan durum wheats grown under rainfed and irrigated conditions. Int J Agron 2021:1–13

Boussakouran A, El Yamani M, Sakar EH, Rharrabti Y (2022) Genetic progress in physiological and biochemical traits related to grain yield in Moroccan durum wheat varieties from 1984 to 2007. Crop Sci 62(1):53–66

Bregaglio S, Frasso N, Pagani V, Stella T, Francone C, Cappelli G, Confalonieri R et al (2015) New multi-model approach gives good estimations of wheat yield under semi-arid climate in Morocco. Agron Sustain Dev 35(1):157–167

Chairi F, Vergara-Diaz O, Vatter T, Aparicio N, Nieto-Taladriz MT, Kefauver SC, Araus JL et al (2018a) Post-green revolution genetic advance in durum wheat: the case of Spain. Field Crop Res 228:158–169

Clarke JM, De Ambrogio E, Hare RA, Roumet P (2012) Genetics and breeding of durum wheat. In: Sissons M, Abecassis J, Marchylo B, Carcea M (eds) Durum Wheat chemistry and technology, 2nd edn. AACC International, St. Paul, MN, USA, pp 15–36

Confalonieri R, Bregaglio S, Cappelli G, Francone C, Carpani M, Acutis M, Dong Q et al (2013) Wheat modeling in Morocco unexpectedly reveals predominance of photosynthesis versus leaf area expansion plant traits. Agron Sustain Dev 33(2):393–403

Cossani CM, Slafer GA, Savin R (2009) Yield and biomass in wheat and barley under a range of conditions in a Mediterranean site. Field Crops Res 112:205–213

Cottrell RS, Nash KL, Halpern BS, Remenyi TA, Corney SP, Fleming A, Blanchard JL et al (2019) Food production shocks across land and sea. Nat Sustain 2(2):130–137

Daszkiewicz T (2022) Food production in the context of global developmental challenges. Agriculture 12(6):832

De Vita P, Mastrangelo AM, Matteu L, Mazzucotelli E, Virzi N, Palumbo M, Cattivelli L et al (2010) Genetic improvement effects on yield stability in durum wheat genotypes grown in Italy. Field Crop Res 119(1):68–77

Del Pozo A, Matus I, Ruf K, Castillo D, Méndez-Espinoza AM, Serret MD (2019) Genetic advance of durum wheat under high yielding conditions: the case of Chile. Agronomy 9:454

Devkota M, Yigezu YA (2020) Explaining yield and gross margin gaps for sustainable intensification of the wheat-based systems in a Mediterranean climate. Agric Syst 185:102946

Dodig D, Zorić M, Kandić V, Perović D, Šurlan-Momirović G (2012) Comparison of responses to drought stress of 100 wheat accessions and landraces to identify opportunities for improving wheat drought resistance. Plant Breed 131(3):369–379

Ehdaie B, Layne AP, Waines JG (2012) Root system plasticity to drought influences grain yield in bread wheat. Euphytica 186(1):219–232

FAOSTAT (Food and Agriculture Organization of the United Nations Statistical Database) (2021) Online database, 2015. FAO, Rome, Italy. http://www.fao.org/faostat/en/#data/QC. Accessed 16 May 2021

Farooq M, Hussain M, Siddique KH (2014) Drought stress in wheat during flowering and grain-filling periods. Crit Rev Plant Sci 33:331–349

Giunta F, Motzo R, Pruneddu G (2007) Trends since 1900 in the yield potential of Italian-bred durum wheat cultivars. Eur J Agron 27:12–24

Jlibene M (2005) Amélioration génétique du blé tendre. Pages 9–55 in La Création Variétale à l'INRA: Méthodologie, Acquis et Perspectives (F. Abbad Andaloussi et A. Chahbar, eds.). INRA, Rabat, Morocco

Kadkol GP, Sissons M (2016) Durum wheat: overview. In: Wrigley C, Corke H, Seetharaman K, Faubion J (eds) Encyclopedia of food grains, 2nd edn. Academic Press, Oxford, pp 117–124

Mackay IJ, Cockram J, Howell P, Powell W (2021) Understanding the classics: the unifying concepts of transgressive segregation, inbreeding depression and heterosis and their central relevance for crop breeding. Plant Biotechnol J 19(1):26–34

MAPMDREF (2019) Ministère de l'Agriculture, de la Pêche Maritime, du Développement Rural et des Eaux et Forêts, Division des statistiques. (http://www.agriculture.gov.ma), Morocco

Mohammadi R (2012) Genetic gain in grain yield and drought tolerance of durum wheat breeding lines under rainfed conditions in Iran. Acta Agron Hung 60(4):417–432

Nsarellah N, Amri A, Miloudi N (2006) Amélioration génétique du blé tendre. In: Abbad Andaloussi F, Chahbar A (eds) La Création Variétale à l'INRA: Méthodologie, Acquis et Perspectives (, eds.). INRA, Rabat, Morocco, pp 59–95

ONSSA (2021) National de Sécurité Sanitaire des produits Alimentaires Available online: 32. http://www.onssa.gov.ma . Accessed on 13 May 2021

Reynolds MP, Mujeeb-Kazi A, Sawkins M (2005) Prospects for utilising plant-adaptive mechanisms to improve wheat and other crops in drought-and salinity-prone environments. Ann. Appl. Biol. 146(2):239–259

Rife TW, Graybosch RA, Poland JA (2019) A field-based analysis of genetic improvement for grain yield in winter wheat cultivars developed in the US central plains from 1992 to 2014. Crop Sci 59(3):905–910

Royo C, Elias EM, Manthey FA (2009) Durum wheat breeding. In: Cereals. Springer, New York, pp 199–226

Ryan J, Pala M, Masri S, Singh M, Harris H (2008) Rainfed wheat-based rotations under Mediterranean conditions: crop sequences, nitrogen fertilization, and stubble grazing in relation to grain and straw quality. Eur J Agron 28(2):112–118

Sanchez-Garcia M, Royo C, Aparicio N, Martin-Sanchez JA, Alvaro F (2013) Genetic improvement of bread wheat yield and associated traits in Spain during the 20th century. J Agric Sci 151:105–118

Teuling AJ, Van Loon AF, Seneviratne SI, Lehner I, Aubinet M, Heinesch B, Spank U et al (2013) Evapotranspiration amplifies European summer drought. Geophys Res Lett 40(10):2071–2075

The International Center for Agriculture Research in the Dry Areas (ICARDA). Available online: https://www.icarda.org/. Accessed on 30 May 2021

Vrochidou AE, Tsanis IK, Grillakis MG, Koutroulis AG (2013) The impact of climate change on hydrometeorological droughts at a basin scale. J Hydrol 476:290–301

Xynias IN, Mylonas I, Korpetis EG, Ninou E, Tsaballa A, Avdikos ID, Mavromatis AG (2020) Durum wheat breeding in the Mediterranean region: Current status and future prospects. Agronomy 10(3):432

Yuan J, Sha ZM, Hassani D, Zhao Z, Cao LK (2017) Assessing environmental impacts of organic and inorganic fertilizer on daily and seasonal greenhouse gases effluxes in rice field. Atmos Environ 155:119–128

Yves T, Koutroulis A, Samaniego L, Vicente-Serrano SM, Volaire F, Boone A, Polcher J (2020) Challenges for drought assessment in the Mediterranean region under future climate scenarios. In: Earth-science reviews, 103348

Chapter 9
Symbiotic Arbuscular Mycorrhizal Fungi as a Soil Essential Component Contributing to Soil Fertility for Sustainable Agriculture in Arid Environment

Lahcen Ouahmane, Hanane Dounas, Mohamed El Yamani, and Kholoud Kahime

Abstract Arid zones are deeply affected by desertification, global warming, and anthropic pressure. Water deficiency and salinity are very worrying problems that negatively affect the growth and the physiological and biochemical activities in crops. In these arid zones, the main challenge is to develop new eco-friendly technologies that improve crop tolerance to abiotic constraints such as drouth and salinity. In this sense, mycorrhizal fungi (MF) have received much attention due to their capability in promoting plant growth and tolerance to abiotic and biotic stresses. It is thus fitting that the current review was undertaken to underline the effects of MF on the growth, the nutrition, the development and the crop tolerance to drouth, salt stress, and their use as biofertilizers. Furthermore, their improvement of the tolerance of crop to pests and diseases by activating the different processes of enzymatic defense and nutritional competition.

Keywords Phosphorus uptake · Water nutrition · Abiotic stressors · Arid zone · Crops

---

L. Ouahmane (✉) · H. Dounas
Laboratory of Microbial Biotechnologies, Agro-Sciences and Environment (BioMAgE), Labeled Research Unit-CNRST N∘4, Cadi Ayyad University, 40000 Marrakesh, Morocco
e-mail: l.ouahmane@uca.ac.ma

M. El Yamani · K. Kahime
Laboratory of Applied Sciences for the Environment and Sustainable Development, Essaouira School of Technology, Cadi Ayyad University, Km 9, Route d'Agadir, BP. 383, Essaouira Al Jadida, Morocco

M. El Yamani
Laboratory of Natural Resources and Environment, Polydisciplinary Faculty of Taza, Sidi Mohamed Ben Abdellah University, BP. 1223, Taza-Gare, Taza, Morocco

K. Kahime et al. (eds.), *Climate Change Effects and Sustainability Needs*, Springer Climate, https://doi.org/10.1007/978-3-031-59603-2_9

Introduction

The term mycorrhiza derives from Greek and refers to the symbiotic relationship between plants and soil fungi. It was first used by the German botanist Frank in 1885 to describe the mixed root-fungus organs. Subsequently, this term evolved into a larger one. Literally, it means "fungus and root" (from the Greek mukês: fungus and rhiza: root). Fungi grow both within and around the roots, forming a network of filaments that greatly amplify the surface area of contact between the roots and the soil (Brundrett 2004). In the mycorrhizal symbiosis, the fungus receives from the host plant sugars and amino acids formed during photosynthesis, while in return the plant receives mineral nutrients from the soil (especially phosphorus) as well as water collected by the fungus (Bago et al. 2000). Due to the extensive mycelium with which the fungus colonizes the soil, the plants have access to a greater proportion of the soil volume, which increases the availability of soil nutrients. Thus, it is believed that a large part of the plant's mineral nutrients come from its fungal partners. This ensures better growth and tolerance to abiotic and biotic stresses like drouth, salinity and diseases (Selosse et al. 2006; Pozo et al. 2010).

Types of Mycorrhizae

Delian et al. (2011) said that there are different types of mycorrhizae based on their ecological, morphological, and physiological characteristics, as well as how they work with plant and fungal species. So far, seven types have been named: superficial interaction (ectomycorrhiza), intermediate ectendomycorrhizae, internal (endomycorrhizae), arbuscular endomycorrhizae, arbutoid, monotropoid, ericoid, and orchid mycorrhizae. These symbiotic associations were recently classified into eight types of mycorrhizae (Fig. 9.1) (Garbaye 2013). Arbuscular and ectomycorrhizal mycorrhizal symbioses are the most abundant and widespread (Bonfante and Genre 2010) (Fig. 9.2). Whereas in the case of Endomycorrhizae (Fig. 9.2), the symbiosis is not visible to the naked eye (Read 1991). The fungi colonize the interior of root cells to form arbuscules and vesicles, both of which play an important role in nutrient exchange with the plant (Bonfante and Genre 2010).

Arbuscular Mycorrhizal Fungi Community

Most mycorrhizal symbioses involve arbuscular mycorrhizal fungi (AMF), which are obligatory symbiont fungi that belong to the single-phyletic phylum Glomerycota. They are characterized by the presence of fungal structures inside the root cells of the host plant. These intracellular structures are highly branched or coiled in the cells of the root cortex (He and Nara 2007), known as arbuscules (from the

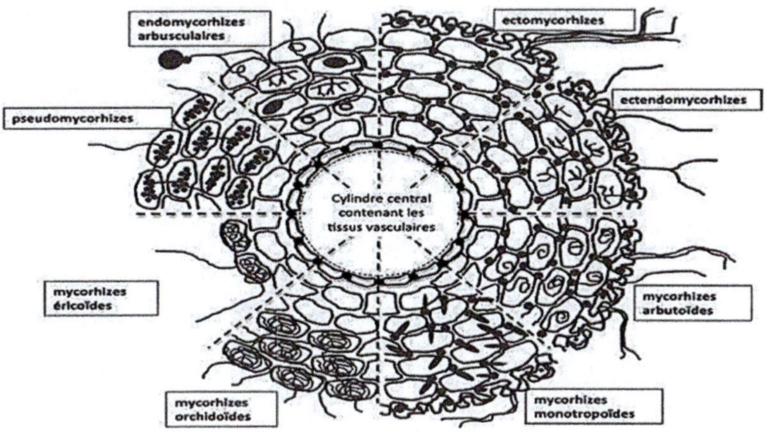

Fig. 9.1 Principal mycorrhizal groups as shown in a root cross section (Garbaye 2013)

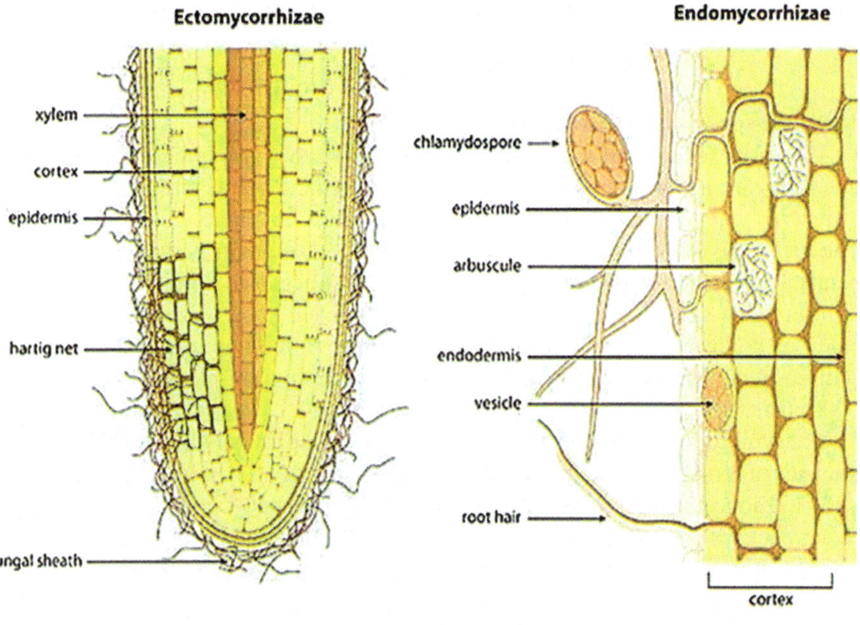

Fig. 9.2 Schematic illustration of the difference between ectomycorrhizae and endomycorrhizae in plant roots (Bonfante and Genre 2010)

Latin arbusculum, which means shrub). Arbuscules are the key feature of symbiosis (AMF), as they represent an extreme of intimacy and compatibility and are considered the site of nutrient exchange between symbiotic partners (Parniske 2008; Garg and Chandel 2011). Among the vascular plants present today, about 80% are considered mycotrophic. Thus, they are found in angiosperms, gymnosperms, pteridophytes,

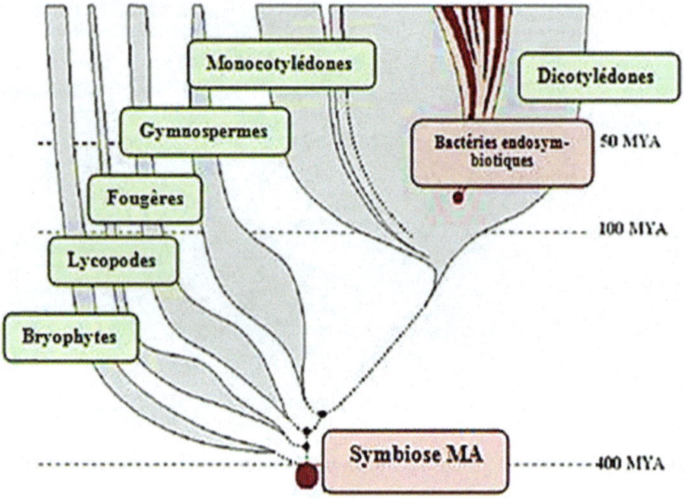

Fig. 9.3 Phylogenetic tree representing the evolution of the two main root symbioses

lycopods, and bryophytes (Hause and Fester 2005). This phylum includes about 250 described species (Baszkowski et al. 2012).

The fossil record and the current distribution of the AM symbiosis suggest that mycorrhization was already present in the earliest land plants and that the relationship was beneficial enough to keep going as vascular plants changed over time. The existence of AM symbiosis dates back to the Devonian period, when the first plants appeared on earth, mainly Bryophytes, and began to colonize the earth 460 million years ago. Findings of hyphae and fossilized spores as well as molecular analyses confirm their ancient existence and lead to the hypothesis that early AM symbiosis contributed significantly to land colonization by plants (Redecker et al. 2000) (Fig. 9.3).

The symbiotic association between arbuscular mycorrhizal fungi (AMF) and most lineages of land plants, including mosses, lycopods, and ferns, originated more than ~400 million years ago (MA). The origin of symbiosis between plants and nitrogen-fixing bacteria is much more recent (~65 MA) (Kistner and Parniske 2002).

Structure of Arbuscular Mycorrhizae and Life Cycle of Endomycorrhizal Fungi

In general, an active mycorrhiza is formed by an extra-root part that includes the external mycelium of the fungus developed in the soil, the spores, the auxiliary cells, an intra-root part that includes intracellular (Paris-type mycorrhizae) and frequently

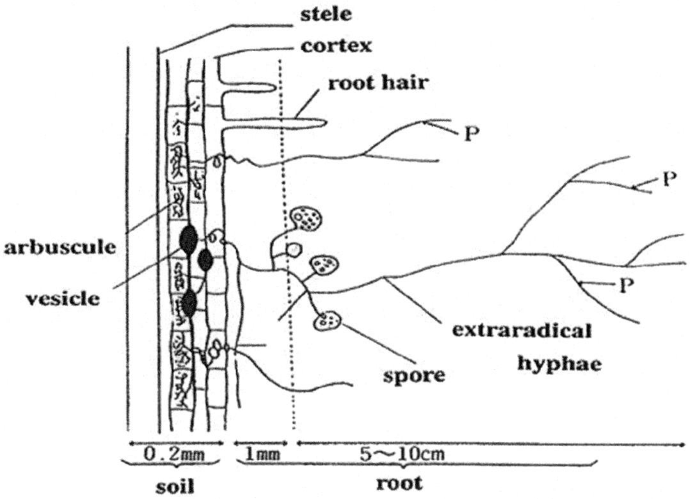

Fig. 9.4 Schematic picture of arbusucular mycorrhizal fungi colonizing roots and their hyphal extension into soil by Takuji Ohyama in (Machi 2014)

intercellular (Arum-type) hyphae; arbuscules, which are formed by repeated dichotomous division of hyphae; and occasionally intracellular (or intercellular) vesicles with a lipid storage function (Fig. 9.4).

In the scientific literature, there are two types of intra-root symbiotic structures: the Arum type and the Paris type. However, there is a continuum of structures in between these two types (Dickson et al. 2007). Arum-type mycorrhizae are better described and more typical, forming intercellular d'hyphae between cortical cells and intracellular arbuscules within cortical cells (Krishnamoorthy et al. 2017), whereas Paris-type mycorrhizae, which are less known but abundant, form large intracellular coils of hyphae and arbuscular coils in the root cortex and not in intracellular spaces (Willis et al. 2013; Krishnamoorthy et al. 2017). It should be noted that when AMF infects a root, it doesn't change its shape in a way that can be seen with the naked eye or by simple macroscopic observation. Also, the fungus only infects the root's epidermis and cortical parenchyma, not the central cylinder or meristematic zones (Bonfante et al. 1984) (Fig. 9.5).

Development Cycle of Arbuscular Mycorrhizal Fungi

The development cycle of AMF can be divided into five main stages (Fig. 9.6). The spores of AM fungi germinate when they are under optimal temperature and humidity conditions. The germ tube of the fungus develops in a limited way with minimal use of reserves. When the fungus perceives the presence of a host plant, it shows a typical reaction of intense branching of the hyphae called "branching". When the fungus

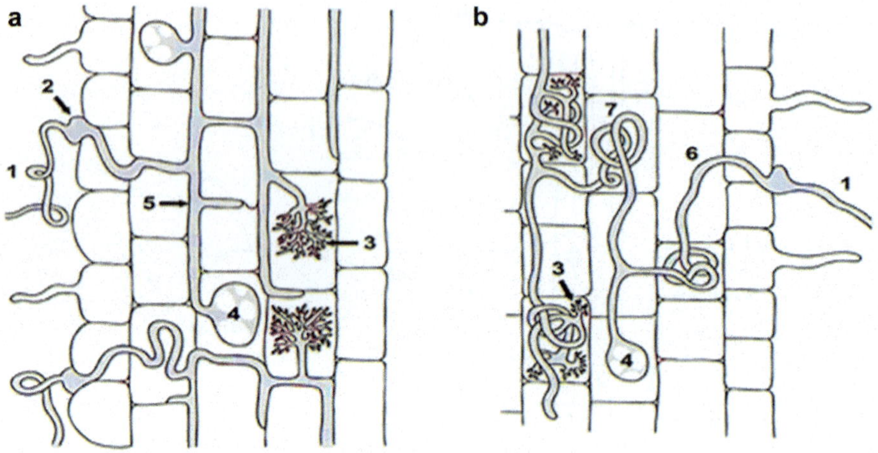

Fig. 9.5 The two main types of root colonization by AMF. **a** Arum type. **b** Paris type.1: extra-root hyphae; 2: appressorium;3: arbuscules; 4: vesicles; 5: intercellular hyphae; 6: intracellular hyphae; 7: peloton (Smith and Read. 2008)

comes into contact with the host root, it differentiates appressoria on the surface of the rhizodermis (blue cells). After apoplastic development, the intraracinar hyphae of the fungus reach the cortical cells (green cells), where they differentiate into highly branched structures called arbuscules. After differentiation of the intraracinar structures, the fungus produces spores from its extra-root mycelium.

Fig. 9.6 Development cycle of AM fungi (Akiyama 2007)

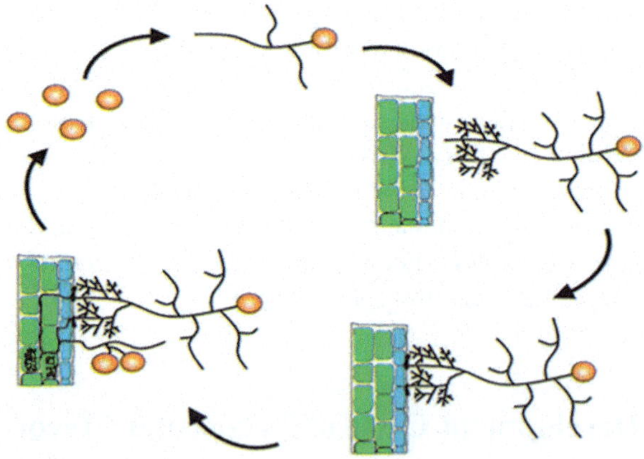

Establishment of the AMF Symbiosis

The AMF does not show any specificity towards the host plant (Krishnamoorthy et al. 2017). Numerous AMF species can infect the same plant, and a fungus can associate with several different plants (Bécard et al. 2004). There are three main sources of inoculation to initiate colonization of a new host root: spores, infected root fragments, or hyphae from an already established mycelium. The colonization of plant roots by the fungus can be summarized in three phases (Bonfante and Genre 2010) (Fig. 9.7).

Asymbiotic Stage

AMF spores can germinate spontaneously under conditions of satisfactory temperature and hygrometry, in the absence of a host plant, and without the need to use their lipid reserves (Bécard et al. 2004). During this phase, the development of the germ tube remains limited. After some time, the protoplasm retracts into the spore, and the spore again enters dormancy. Some spores may be able to germinate several times (Koske 1981; Giovannetti et al. 2002), which leads to the suggestion that during the asymbiotic phase, the spores retain their growth potential and ability to survive (Logi et al. 1998; Giovannetti 2001). This symbiosis requires a high level of coordination between the two partners.

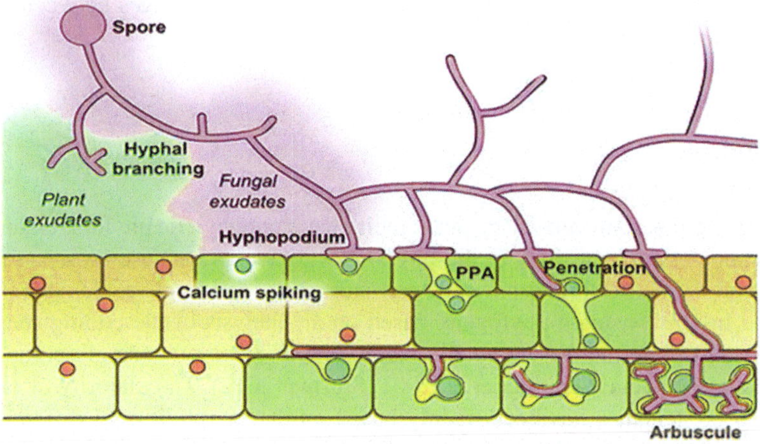

Fig. 9.7 Schematic presentation of AMF root colonization process by (Bonfante and Genre 2010)

Pre-Symbiotic Stage

This stage involves a prior molecular dialogue between the future partners in the rhizosphere. The perception of diffuse signals by both partners allows them to detect each other's presence before any physical contact and leads to the development of fungal structures (Bonfante and Genre 2010). In the pre-contact stage, mutual recognition is characterized by hyphal branching responses induced by plant-derived strigolactones (SLs) (Fig. 9.8) (Akiyama 2007) and plant gene expression induced by diffuse fungal signal molecules: the Myc factors (Fig. 9.7) (Kosuta et al. 2003; Kuhn et al. 2010).

This phase is characterized by abundant branching of the germ tube, increasing the probability of contact with the roots and thus the establishment of symbiosis (Akiyama 2007; Besserer et al. 2006).

Symbiotic Phase

When the hyphae of the AMF come in contact with the root, they round out and flatten on the root wall to form a structure called a "hyphopodium" or "appresorium." These are structures that attach to and penetrate the cells of the epidermis. The fungal hyphae can penetrate the root cells to the inner cortical cells, where the fungus forms intracellular structures and produces vesicles involved in the storage of lipids and arbuscules that remain separated from the cytoplasm of the cells by an extension derived from the plasma membrane, the periarbuscular membrane that surrounds the hyphal branches, constituting the site of nutritional exchanges between the two partners (Smith and Read 2011).

Taxonomy

Before the glomeromycetes were put in their own monophyletic phylum, Gerdemann and Trappe (1974) put them in the phylum Zygomycota because they had similar spore shapes, spore formation, cell wall structures, and hyphae. Schübler finally showed in 2001, using a phylogeny based on an analysis of the coding sequences of the small ribosomal RNA subunit, that the AMF belong to a single group called the Glomeromycota (Schüßler et al. 2001; Stürmer 2012). The majority of research on AMF systematics has been mainly based on the morphological differences of spores isolated from soil (spore ontogeny, color, shape, ornamentation, wall, etc.) (Ruissen 2013). And which play a very important role in the identification, description, and classification of new AMF species. However, the variable morphological characters, the obligate biotrophic lifestyle, and the absence of sexual reproduction in AMF make species identification difficult and complex (Sanders et al. 1996;

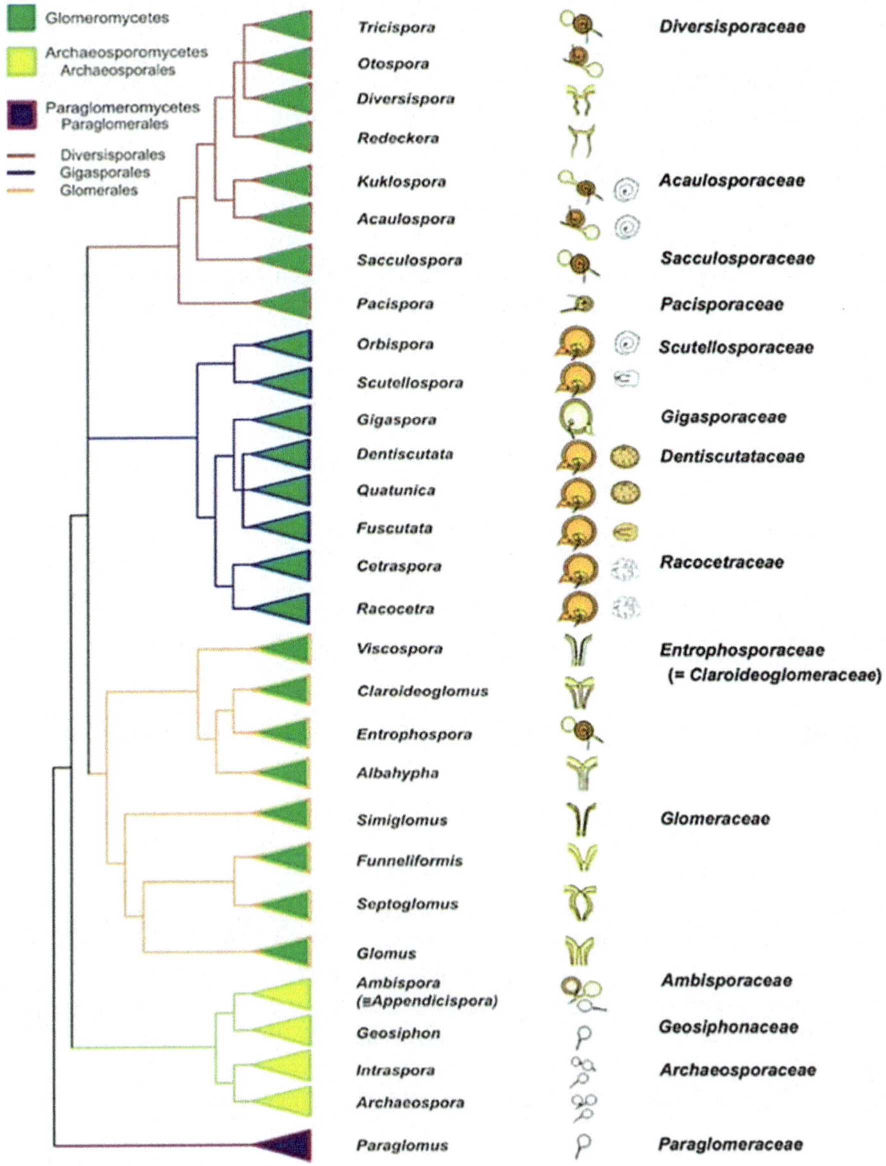

Fig. 9.8 Representative tree of the phylum Glomeromycota on molecular basis (SSU, ITS region, partial LSU gene rRNA, and partial β-tubulin gene) and morphological analyses (Oehl et al. 2011)

Dickson 2004; Redecker and Raab 2006). The classification of this group of fungi has undergone considerable transformation and has been newly structured based on the combined analysis of genetic and phenotypic characteristics (Oehl et al. 2011). Stürmer (2012) discusses the history of taxonomy by highlighting four time periods. The "initial discovery" period (1845–1974) is characterized by descriptions, especially of sporocarp-forming species. The "alpha taxonomy" period (1975–1989) created a solid morphological basis for species identification and classification. The cladistic period (1990–2000) saw the first cladistic classification of AMF based on phenotypic characters only. By the end of this period, genetic characters had played a role in defining taxa and elucidating evolutionary relationships within the group. The period of phylogenetic synthesis (2001–present) began with the proposal of a new classification based on genetic characters. Currently, and with the help of the molecular biology tool, phylogenetic studies based on SSU rDNA sequences have led to the separation of the Glomales into a monophyletic phylum: Glomeromycota (Schüßler et al. 2001; Ruissen 2013). This phylum is divided into three classes: Archaeosporomycetes, Glomeromycetes, and Paraglomeromycetes, with five orders: Glomerales, Diversisporales, Gigasporales, Archaeosporales, and Paraglomerales (Fig. 9.8). Fifteen families, with 38 genera and approximately 270 species, are known to date (Oehl et al. 2011; Goto et al. 2012; Blaszkowski et al. 2015; Marinho et al. 2014).

Benefits of Arbuscular Mycorrhizal Symbiosis

AMF are considered key players not only in the dynamics and productivity of the ecosystem but also in the processes ensuring the coexistence of plants (Hart et al. 2003; Van Der Heijden et al. 1998; Oliveira et al. 2006). It is an essential element in the life cycle of the host plant (growth and reproduction) (Smith and Read 2011; Begum et al. 2019). Mycorrhizal symbiosis is strongly governed by the reciprocal exchanges between the two partners. Indeed, the impact of mycorrhizal fungi on the host plant is not limited to an improvement of hydro-mineral nutrition but also in the enhancement of the tolerance of the host plants to different biotic and abiotic stresses, in the structuring of the soil, and in the improvement of the soil fertility (Qiangsheng and Renxue 2006; Rillig and Mummey 2006; Xu et al. 2018).

Improvement of Plant Growth and Nutrition

The AMF play an important role in improving vegetative growth and coexistence of plants on very difficult conditions. Indirectly through a natural strategy of nutrient and water acquisition (Finlay 2004; Smith and Read 2011) AMF play an important role in the uptake of nutrients, especially those of low mobility in the soil such as phosphorus (P) (Berruti et al. 2016; Wahid et al. 2019) by an external hyphal network that functions as an extension of the host plant's roots, improving its efficiency in exploring the soil thus increasing the surface area for exchange between the plant and its environment (Hodge et al. 2010).

Nutrient Exchange

Currently, it is well known that the primary role of AMF is to increase the uptake of poorly mobile elements in the soil during the interaction; the plant provides the fungus with carbon compounds from photosynthesis (Bago et al. 2003). The extra-radicular mycelium is unable to take the necessary carbon compounds from the environment, and it is therefore the intra-radicular mycelium that absorbs almost all the carbon compounds necessary for the fungus from the plant (Wiemken 2007). Thus, in the fungus, there is a flow of minerals from the extra-radicular mycelium to the plant and a flow of carbon from the root to the extra-radicular mycelium (Godfroy 2008). The fungus provides the plant with mineral compounds such as potassium, calcium, magnesium, copper, nitrogen, water, and especially phosphorus. (Fig. 9.10) (Ramos et al. 2011; Kaya et al. 2009; Richardson et al. 2009; Farzaneh et al. 2011). The fungus collects the minerals from the soil and provides them to the plant, with the extra-radicular mycelial network constituting an interface area with the soil. The three-dimensional extent of this network allows the exploration of a considerably larger volume of soil than that accessible by the host plant's root system alone. Sylvia (1986) measured an average of 12 m of AM fungal hyphae per gram of soil (Sylvia 1986) (Fig. 9.9).

Phosphorus Uptake

Phosphorus (P) is an essential element classified as a macronutrient due to the relatively large amounts required by plants. Plant growth is related to the availability of inorganic phosphorus in the plant rhizosphere (Holford 1997). Phosphorus is an essential component of adenosine triphosphate (ATP) in nucleic acids and membrane phospholipids (Marschner 1995; Gaude et al. 2008; Maathuis 2009). It assists in photosynthesis, energy conservation, carbon metabolism, redox reactions, enzyme activation and inactivation, signaling, and nucleic acid synthesis (Vance et al. 2003). (Vance et al. 2003; Gaude et al. 2008; Maathuis 2009; Ramos et al. 2011). In general, P availability in the soil is very low due to its fixation and immobilization (Javot et al. 2007; Yadav and Tarafdar 2010). Plants can take up P directly via the soil-root interface through the epidermis and root hairs. Under the action of root uptake, depleted zones are rapidly created around the roots. A small proportion (usually less than 1%) is immediately available to plants, which have difficulty acquiring this element while their needs are great (Richardson et al. 2009; Smith and Smith 2011). To access soil P pools inaccessible to plants, another strategy is termed the "mycorrhizal" pathway,

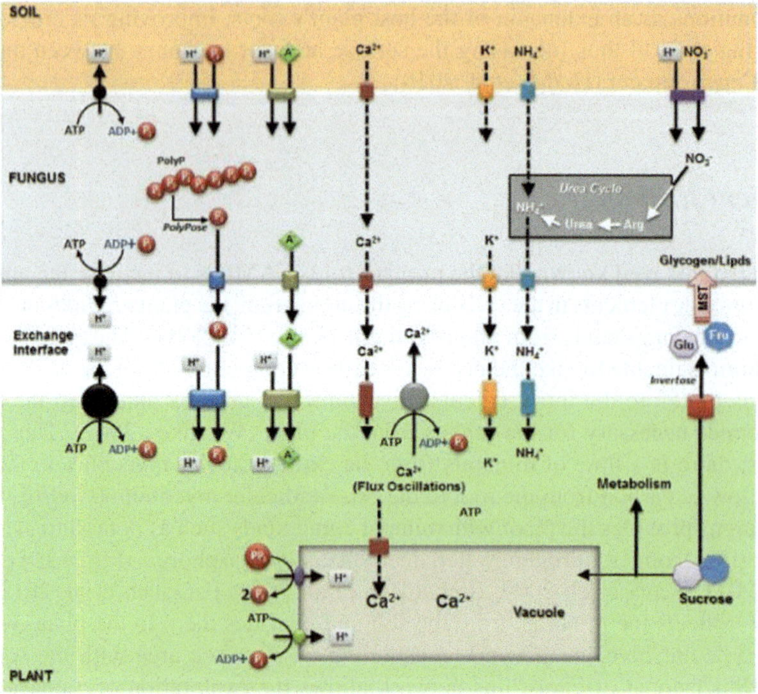

Fig. 9.9 Schematic representation of the mechanisms involved in the exchange of major nutrients between the AMF and host plant (Ramos et al. 2011)

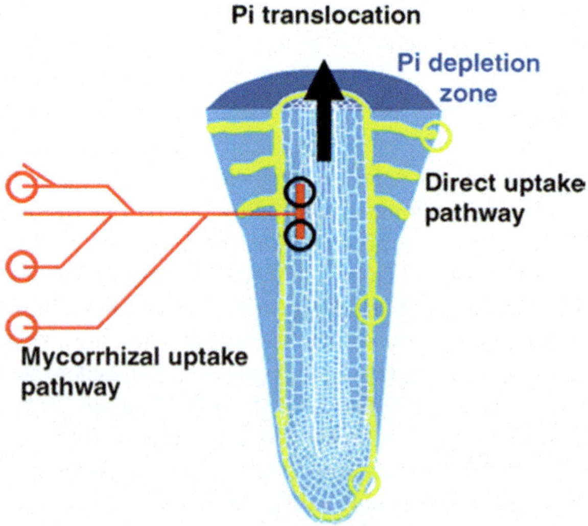

Fig. 9.10 Schematic representation of the two pathways of phosphorus uptake in a root. (Bucher 2007)

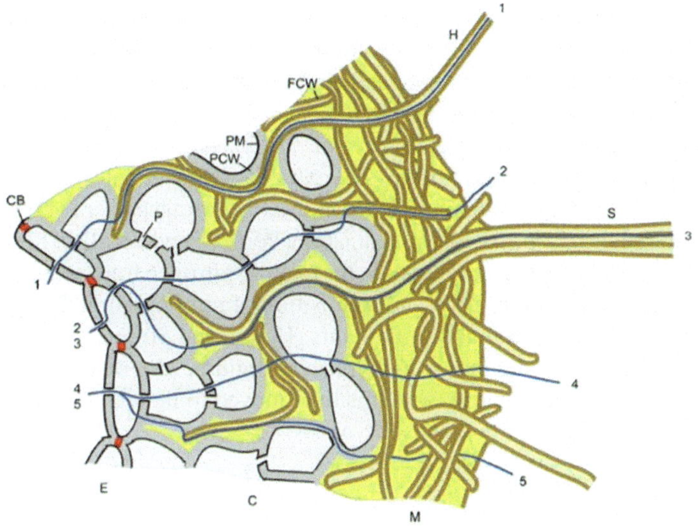

Fig. 9.11 Possible pathways for water transport between soil and root xylem in an ectomycorrhizal root of a gymnosperm. FCW, fungal cell wall; PCW, plant cell wall; PM, plant plasma membrane; P, plasmodesm; CB, casparian band; E, endoderm; C, net area of cortex/Hartig; M, mantle; S, strand; H, hypha. The hypal membrane is not shown separately. The yellow matrix is filled with hyphae and their cell walls

via the extra-root mycelium (Fig. 9.11) (Karandashov and Bucher 2005; Bucher 2007; Smith and Smith 2011). The AMF would be able to hydrolyze organic P into inorganic P to make it available in the soil to the plant or transfer it directly to the host plant (Feng et al. 2003; Shibata and Yano 2003).

a. Direct pathway, Pi transporters in the plant located in the epidermis, have the role of absorbing Pi from the soil directly into the root cells. (Yellow circles) formation of a depletionzone (bluezone) near the root surface.
b. Indirect symbiotic pathway involves AMF taking up Pi by fungal Pi transporters (red circles) located in the extra-matrix hyphae. Pi is the n transferred to the root and eventually to the symbiotic interface in the cortex. The Pht1transporters in the plant, located in this interface (black circles).

Nitrogen Nutrition

Nitrogen is an essential and indispensable element for plant growth like phosphorus (Ohyama 2010) and contributes to the structure and synthesis of many molecules such as amino acids, coenzymes, nucleotides, and phospholipids (Hawkesford et al. 2012). It has been shown that mycorrhizal roots can access important sources of nitrogen (Govindarajulu et al. 2005; Hodge et al. 2010; Hodge and Storer 2015).

Nitrogen transfer by AMF has been demonstrated using radioactive isotopes (Govindarajulu et al. 2005; Jin et al. 2005; Miransari 2011). The AMF possesses the enzymatic equipment necessary for nitrate and ammonium acquisition. They can transfer significant amounts of inorganic nitrogen to their host plants (Govindarajulu et al. 2005; Hodge et al. 2010; Hawkesford et al. 2012). Nitrogen uptake requires the activity of transporters localized at the soil/extra-root hyphae interface of the AMF. GintAMTI and GintAMT2, which code for high-affinity transporters of NH4+ ions in Glomus intraradices (López-Pedrosa et al. 2006; Pérez-Tienda et al. 2011), have been found in AMF. They are both responsible for making nitrogen transporters. AMF can also accelerate the decomposition and acquisition of nitrogen directly from organic matter to increase its bioavailability to plants (Hodge and Storer 2015). Once taken in, nitrogen is moved to the intra-root hyphae as arginine (Jin et al. 2005). There, it is broken down into ammonium, which is then moved to the plant at the arbuscular level (He et al. 2003).

Enhancement of the Uptake of Other Mineral Elements

Generally, AMF also improves plant uptake of other mineral elements such as K, Ca, and Mg (Clark and Zeto 2000; Farzaneh et al. 2011; Ramos et al. 2011). Using the extramural network lets the plant look at a larger area of soil and get other trace elements that aren't easily moved, like S (Allen and Shachar-Hill 2009), Cu, Zn, Mn, and Fe (Liu et al. 2000; Lehmann and Rillig 2015; Bagheri et al. 2018). However, when these elements are present in high quantities and therefore have a polluting character, the AMF can play a protective role for the plant by retaining these toxic elements (Kumar and Saxena 2019).

Improvement of Water Nutrition

The lack of nutrients in arid soils and seedlings and the long dry periods that can last for several months restrict plant nutrient and water resources. These stressful conditions force plants to seek these resources far from the rhizosphere, with the help of hyphae from the mycorrhizal symbiotic association. Püschel et al. (2020) showed that adding the AM fungus Rhizophagusirregularis to the Medicago truncatula plant made it take in more water. Zarik et al. (2016) showed that inoculating Cupressus atlantica G with arbuscular mycorrhizal fungi made it more resistant to drought. This happened because the extended extraracinar hyphae of AM fungi helped the plant take in more water. For their part, Lehto et al. (2011) demonstrated that ectomycorrhizae play an important role in water uptake by plants. Figure 9.11 shows the possible pathways for water transport between soil and root xylem in an ectomycorrhizal root of a gymnosperm. The models representing the possible pathways for water transport between the soil and the root xylem in an ectomycorrhizal root of a gymnosperm consist of the following steps:

(a) Water enters an individual hypha in the soil, is carried into the fungal symplast, and then moves to the root symplast through the shared apoplast;
(b) Same as (1) except that water penetrates the symplast of the hypha in the mantle;
(c) Water enters the apoplast of a mycelial strand, is transported into the fungal and plant apoplasts, and then enters the root cell;
(d) trans membrane pathway: water passes through several fungal and plant membranes, then into the plant symplast;
(e) Water moves through the fungal and plant apoplasts to the endodermis, where the Casparian band forces it into the symplast (Lehto and Zwiazek 2011).

Protection Against Abiotic and Biotic Stresses

It is well known that the main function of AMF is the improvement of water and mineral nutrition in plants. Another important characteristic of AMF is its protective effect on the host plant by improving tolerance against abiotic and biotic stresses such as salinity (Borde et al. 2017), drought (Ruiz-Sánchez et al. 2010), and temperature (Latef and Chaoxing 2011a). They are also involved in mitigating the harmful effects of pollutants such as polycyclic aromatic hydrocarbons (Lenoir et al. 2017), fungicides (Campagnac et al. 2010), and trace metals (Abu-Elsaoud et al. 2017; Fall et al. 2021; Khalid et al. 2021). Similarly, AMF is involved in improving resistance to certain cryptogamic diseases (Krzyzaniak et al. 2021). AM fungi can reduce stress symptoms by complementing the plant's intrinsic protective mechanisms (Symanczik et al. 2020). Several studies have shown the role played by AMF in the protection of plants under abiotic stresses (Begum et al. 2019).

Drought Stress

Drought is one of the main stresses that can significantly reduce plant productivity (Farooq et al. 2009). When there isn't enough water in the root zone, plants usually have a high transpiration rate, which closes the stomata and reduces the amount of $CO2$ that gets in. This makes photosynthesis less active (Pagter et al. 2005) and speeds up the production of ROS, which causes oxidative stress in plants (Hasanuzzaman et al. 2013). AMF has been shown to improve plant performance to reduce stress symptoms by complementing intrinsic plant protective mechanisms by improving root length, leaf area, plant biomass, and nutrient uptake under drought conditions. (Medina 2010; Mathur et al. 2019). Arbuscular mycorrhizal fungi improve the water status of the host plant in several ways that can be linked to better water and nutrient uptake in the mycorrhizal roots. Extra-radical hyphae and the release of glomalins make soil aggregates much more stable (Ji et al. 2019; Chen et al. 2020). AMF-activated microbial life has a significant impact on aggregate stability in a semi-arid, degraded area. which in turn improves water and nutrient uptake as well as soil structure (Pagano 2014; Ji et al. 2019), and AMF improves and regulates photosynthetic rate. In general, the higher photosynthetic rate in mycorrhizal plants leads to higher carbohydrate production. It has been demonstrated that mycorrhized barley plants under water stress have carbohydrate concentrations that are about 80% higher than those of non-mycorrhized plants (Porcel and Ruiz-Lozano 2004; Khalvati et al. 2005). In addition to these benefits, the AMF quickly activates a host plant defense system. This protects the host plant from oxidative damage caused by water stress and makes it make antioxidant enzymes like superoxide dismutase, catalase, and guaiacolperoxidase, as well as non-enzymatic antioxidant molecules like ascorbate and glutathione (Ruiz-Sánchez et al. 2010; Liu et al. 2017).

Salinity Stress

Salinity is now one of the most important abiotic stresses. It slows growth and development, changes photosynthesis (Raziuddin et al. 2011), and makes more reactive oxygen species (ROS), which affect lipid and protein metabolism (Hasanuzzaman et al. 2013; Hashem et al. 2014; Ahmad et al. 2016). Many authors have attributed the improvement of plant growth and yield under salinity conditions by amf to improved water and nutrient uptake and increased photosynthesis (Latef and Chaoxing 2011b; Beltrano et al. 2013; Ahmad et al. 2016). It has been shown that colonization by AMF allows salinity-exposed host plants through enhanced water uptake through their hyphal network and to improve gas exchange (Porcel et al. 2012; Hameed et al. 2014), and can also improve root hydraulic conductivity at low water potential (Elhindi et al. 2017; Bitterlich et al. 2018) and through stimulation and modification of root system morphology (Fan et al. 2011; Qin et al. 2021) and increased stomatal conductance (Zhu et al. 2018). It has been reported that mycorrhizal plants under saline conditions exhibited increased chlorophyll content, greater N and Mg uptake, and inhibition of Na transport compared to non-inoculated plants (Borde et al. 2017; Bencherif et al. 2019; Kong et al. 2020). The main role of AMF is the attenuation of salinity-related oxidative stress damage by increasing the activity of antioxidant enzymes such as SOD, CAT, PPO, and POD or by increasing the concentration of non-enzymatic antioxidant molecules such as ascorbate and glutathione (Ye et al. 2019; Qin et al. 2021). Inoculation with AMF showed higher SOD, POD, CAT, and POX activity in several plants (Dastogeer et al. 2020; Qin et al. 2021). Indeed, AMF possess various oxidative stress-related genes (SOD genes) that, through upregulation, allow mycorrhized plants to tolerate environmental stresses such as drought and salinity (Huang et al. 2008; Ye et al. 2019).

Mycorrhizal Fungi as Bioprotective Agents

Pathogenic elements can also disrupt plants, and the various interactions between nature's components can be advantageous or harmful in their environment, particularly the soil. To protect themselves against these threats, plants have developed protective systems, including the establishment of the mycorrhizal symbiotic association, which can predispose the plant to respond more rapidly to pathogenic attacks (Harrier and Watson 2004). The use of mycorrhizae in agriculture should therefore be seen as a proactive strategy for biological control. According to a lot of research done so far, the use of arbuscular mycorrhizae (AM) is a promising, easy-to-use, and environmentally friendly method of prevention. The presence of AM also makes it much less important to use chemicals, fertilizers, and pesticides from the outside (Dalpé 2005). Among the researchers who have carried out work in this direction, Reyes-tenaet al. (2017) have shown the existence of an associative effect with co-inoculation by mycorrhizal fungi and actinomycetes in the promotion of plant growth and bioprotection against wilt caused by Phytophthora capsici in pepper trees. For their part, Minchin et al. (2012) have shown that ectomycorrhizae (ECM) have a positive effect on seedling performance due to the beneficial relationship between plants and mycorrhizal fungi. They are also effective against various plant root rot diseases. Numerous studies have demonstrated the protective role of ECM not only against fungal pathogens (Morin et al. 1999) but also against nematodes. Furthermore have demonstrated the importance of ECM symbiosis in controlling pine wilt disease, which is a worldwide threat.

Conclusion

Mycorrhizal fungi are key organisms in soils that affect biogeochemical cycles of major crop nutrients. Their role in soil fertility is critical due to their numerous effects on soil composition and microflora. Fungal hyphae have a considerable impact on the soil because they deposit water and minerals that the plant cannot use, thereby playing an important role in sustaining soil fertility, particularly in arid environments with multiple abiotic constraints.

References

Abu-Elsaoud AM, Nafady NA, Abdel-Azeem AM (2017) Arbuscular mycorrhizal strategy for zinc mycoremediation and diminished translocation to shoots and grains in wheat. PLoS ONE 12:e0188220

Ahmad P, Abdel Latef AA, Hashem A, AbdAllah EF, Gucel S, Tran L-SP (2016) Nitric oxide mitigates salt stress by regulating levels of osmolytes and antioxidant enzymes in chickpea. Front Plant Sci 7:347

Akiyama K (2007) Chemical identification and functional analysis of apocarotenoids involved in the development of arbuscular mycorrhizal symbiosis. Biosci Biotechnol Biochem 71:1405–1414. https://doi.org/10.1271/bbb.70023

Allen JW, Shachar-Hill Y (2009) Sulfur transfer through an arbuscular mycorrhiza. Plant Physiol 149:549–560

Bagheri V, Shamshiri MH, Shirani H, Roosta HR (2018) Nutrient uptake and distribution in mycorrhizal pistachio seedlings under drought stress. J Agr Sci Tech 14:1591–1604

Bago B, Pfeffer PE, Abubaker J, Jun J, Allen JW, Brouillette J, Douds DD, Lammers PJ, Shachar-Hill Y (2003) Carbon export from arbuscular mycorrhizal roots involves the translocation of carbohydrate as well as lipid. Plant Physiol 131:1496–1507

Bago B, Pfeffer PE, Shachar-Hill Y (2000) Carbon metabolism and transport in arbuscular mycorrhizas. Plant Physiol 124(3):949–958. https://doi.org/10.1104/pp.124.3.949

Bécard G, Kosuta S, Tamasloukht M, Séjalon-Delmas N, Roux C (2004) Partner communication in the arbuscular mycorrhizal interaction. Can J Bot 82:1186–1197

Begum N, Qin C, Ahanger MA, Raza S, Khan MI, Ashraf M, Ahmed N, Zhang L (2019) Role of arbuscular mycorrhizal fungi in plant growth regulation: implications in abiotic stress tolerance. Front Plant Sci 10:1–15. https://doi.org/10.3389/fpls.2019.01068

Beltrano J, Ruscitti M, Arango MC, Ronco M (2013) Effects of arbuscular mycorrhiza inoculation on plant growth, biological and physiological parameters and mineral nutrition in pepper grown under different salinity and p levels. J Soil Sci Plant Nutr 13:123–141. https://doi.org/10.4067/s0718-95162013005000012

Bencherif K, Dalpé Y, Hadj-Sahraoui AL (2019) Arbuscular mycorrhizal fungi alleviate soil salinity stress in arid and semiarid areas. In: Microorganisms in saline environments: strategies and functions. Springer, Heidelberg, pp 375–400

Berruti A, Lumini E, Balestrini R, Bianciotto V (2016) Arbuscular mycorrhizal fungi as natural biofertilizers: let's benefit from past successes. Front Microbiol 6:1559

Besserer A, Puech-Pagès V, Kiefer P, Gomez-Roldan V, Jauneau A, Roy S, Portais J-C, Roux C, Bécard G, Séjalon-Delmas N (2006) Strigolactones stimulate arbuscular mycorrhizal fungi by activating mitochondria. PLoS Biol 4:e226

Bitterlich M, Franken P, Graefe J (2018) Arbuscular mycorrhiza improves substrate hydraulic conductivity in the plant available moisture range under root growth exclusion. Front Plant Sci 9:301

Błaszkowski J, Chwat G, Góralska A, Ryszka P, Kovács GM (2015) Two new genera, Dominikia and Kamienskia, and D. disticha sp. nov. in Glomeromycota. Nova Hedwigia 100(1–2):225–238

Błaszkowski J, Kovács GM, Gáspár BK, Balázs TK, Buscot F, Ryszka P (2012) The arbuscular mycorrhizal Paraglomus majewskii sp. nov. represents a distinct basal lineage in Glomeromycota. Mycorrhiza 104(1):148–156. doi: https://doi.org/10.3852/10-430

Bonfante P, Genre A (2010) Mechanisms underlying beneficial plant—fungus interactions in mycorrhizal symbiosis. Nat Commun 1:1–11. https://doi.org/10.1038/ncomms1046

Bonfante-Fasolo P, Vian B (1984) Wall texture in the spore of a vesicular-arbuscular mycorrhizal fungus. Protoplasma 120:51–60

Borde M, Dudhane M, Kulkarni M (2017) Role of arbuscular mycorrhizal fungi (AMF) in salinity tolerance and growth response in plants under salt stress condition. In: Mycorrhiza-eco-physiology, secondary metabolites, nanomaterials. Springer, Heidelberg, pp 71–86

Brundrett M (2004) Diversity and classification of mycorrhizal associations. Biol Rev

Bucher M (2007) Functional biology of plant phosphate uptake at root and mycorrhiza interfaces. New Phytol 173:11–26

Campagnac E, Sahraoui AL-H, Debiane D, Fontaine J, Laruelle F, Garçon G, Verdin A, Durand R, Shirali P, Grandmougin-Ferjani A (2010) Arbuscular mycorrhiza partially protect chicory roots against oxidative stress induced by two fungicides, fenpropimorph and fenhexamid. Mycorrhiza 20:167–178

Chen W, Meng P, Feng H, Wang C (2020) Effects of arbuscular mycorrhizal fungi on growth and physiological performance of Catalpa bungei CA Mey. under drought stress. Forests 11.10 1117

Clark RB, Zeto SK (2000) Mineral acquisition by arbuscular mycorrhizal plants. J Plant Nutr 23:867–902

Dalpé Y (2005) Les Mycorhizes: Un Outil de Protection Des Plantes Mais Non Une Panacée. Phytoprotection 86(1):53–59. https://doi.org/10.7202/011715ar

Dastogeer KMG, Zahan MI, Tahjib-Ul-Arif M, Akter MA, Okazaki S (2020) Plant salinity tolerance conferred by arbuscular mycorrhizal fungi and associated mechanisms: a meta-analysis. Front Plant Sci 11

Delian E, Chira A, ChiraL, Savulescu E (2011) Arbuscular mycorrhizae: an overview. South West. J Hort Biol Env 2:167–192

Dickson S (2004) The Arum–Paris continuum of mycorrhizal symbioses. New Phytol 163:187–200

Dickson S, Smith FA, Smith SE (2007) Structural differences in arbuscular mycorrhizal symbioses: more than 100 years after Gallaud, where next? Mycorrhiza 17:375–393

Elhindi KM, El-Din AS, Elgorban AM (2017) The impact of arbuscular mycorrhizal fungi in mitigating salt-induced adverse effects in sweet basil (*Ocimum basilicum* L.). Saudi J Biol Sci 24:170–179

Fall AF, Nakabonge G, Ssekandi J, Founoune-Mboup H, Apori SO, Ndiaye A, Badji A, Ngom K (2021) Roles of arbuscular mycorrhizal fungi on soil fertility and its importance in phytoremediation of polluted soil. Preprints, 2021090284. https://doi.org/10.20944/preprints202109.0284.v1

Fan L, Dalpé Y, Fang C, Dubé C, Khanizadeh S (2011) Influence of arbuscular mycorrhizae on biomass and root morphology of selected strawberry cultivars under salt stress. Botany 89:397–403

Farooq M, Wahid A, Kobayashi N, Fujita D, Basra SMA (2009) Plant drought stress: effects, mechanisms and management. Sustain Agric 153–188

Farzaneh M, Vierheilig H, Lössll A, Kaull HP (2011) Arbuscular mycorrhiza enhances nutrient uptake in chickpea. Plant Soil Environ 57:465–470. https://doi.org/10.17221/133/2011-pse

Feng G, Song YC, Li XL, Christie P (2003) Contribution of arbuscular mycorrhizal fungi to utilization of organic sources of phosphorus by red clover in a calcareous soil. Appl Soil Ecol 22:139–148

Finlay RD (2004) Mycorrhizal fungi and their multifunctional roles. Mycologist 18:91–96

Garbaye J (2013) La symbiose mycorhizienne: une association entre les plantes et les champignons. Editions Quae

Garg N, Chandel S (2011) Arbuscular mycorrhizal networks: process and functions. In: Lichtfouse E, Hamelin M, Navarrete M, Debaeke P (eds) Sustainable agriculture, vol 2. Springer, Dordrecht. https://doi.org/10.1007/978-94-007-0394-0_40

Gaude N, Nakamura Y, Scheible W, Ohta H, Dörmann P (2008) Phospholipase C5 (NPC5) is involved in galactolipid accumulation during phosphate limitation in leaves of Arabidopsis. Plant J 56:28–39

Gerdemann JW, Trappe JM (1974) The endogonaceae in the Pacific Northwest. Mycologia Mmoir N° 5

Giovannetti M (2001) Survival strategies in arbuscular mycorrhizal symbionts. In: Symbiosis. Springer, Heidelberg, pp 293–307

Giovannetti M, Sbrana C, Avio L (2002) Arbuscular mycorrhizal fungal mycelium: from germlings to hyphal networks. In: Mycorrhizal technology in agriculture. Springer, Heidelberg, pp 49–58

Godfroy O (2008) Etude génétique et moléculaire de deux gènes de *Medicago truncatula*, DMI3 et RPG, contrôlant l'établissement de symbioses racinaires. Thèse de doctorat en Biosciences végétales. Toulouse Paul Sabatier University

Goto BT, Maia LC (2012) Glomerospores: a new denomination for the spores of Glomeromycota, a group molecularly distinct from the Zygomycota. Mycotaxon 96(4):129–132

Govindarajulu M, Pfeffer PE, Jin H, Abubaker J, Douds DD, Allen JW, Bücking H, Lammers PJ, Shachar-Hill Y (2005) Nitrogen transfer in the arbuscular mycorrhizal symbiosis. Nature 435:819–823

Hameed A, Dilfuza E, Abd-Allah EF, Hashem A, Kumar A, Ahmad P (2014) Salinity stress and arbuscular mycorrhizal symbiosis in plants. In: Use of microbes for the alleviation of soil stresses, vol 1. Springer, Heidelberg, pp 139–159

Harrier LA, Watson CA (2004) The potential role of arbuscular mycorrhizal (AM) fungi in the bioprotection of plants against soil-borne pathogens in organic and/or other sustainable farming systems. Pest Manag Sci 60:149–157. https://doi.org/10.1002/ps.820

Hart MM, Reader RJ, Klironomos JN (2003) Plant coexistence mediated by arbuscular mycorrhizal fungi. Trends Ecol Evol 18:418–423. https://doi.org/10.1016/S0169-5347(03)00127-7

Hasanuzzaman M, Nahar K, Gill SS, Fujita M (2013) Drought stress responses in plants, oxidative stress, and antioxidant defense. Clim Chang Plant Abiotic Stress Toler 209–250

Hashem A, Abd Allah EF, Alqarawi AA, Al Huqail AA, Egamberdieva D (2014) Alleviation of abiotic salt stress in *Ochradenus baccatus* (Del.) by *Trichoderma hamatum* (Bonord.) Bainier. J Plant Interact 9:857–868

Hause B, Fester T (2005) Molecular and cell biology of arbuscular mycorrhizal symbiosis. Planta 221:184–196. https://doi.org/10.1007/s00425-004-1436-x

Hawkesford M, Horst W, Kichey T, Lambers H, Schjoerring J, Møller IS, White P (2012) Functions of macronutrients. In: Marschner's mineral nutrition of higher plants. Elsevier, pp 35–189

He X, Nara K (2007) Element biofortification: can mycorrhizas potentially offer a more effective and sustainable pathway to curb human malnutrition? Trends Plant Sci 12:331–333

He X-H, Critchley C, Bledsoe C (2003) Nitrogen transfer within and between plants through common mycorrhizal networks (CMNs). CRC Crit Rev Plant Sci 22:531–567

Hodge A, Helgason T, Fitter AH (2010) Nutritional ecology of arbuscular mycorrhizal fungi. Fungal Ecol 3:267–273

Hodge A, Storer K (2015) Arbuscular mycorrhiza and nitrogen: implications for individual plants through to ecosystems. Plant Soil 386:1–19

Holford ICR (1997) Soil phosphorus: its measurement, and its uptake by plants. Aust J Soil Res 1997(35):227–239

Huang LL, Yang C, Zhao Y, Xu X, Xu Q, Li GZ, Cao J, Herbert SJ, Hao L (2008) Antioxidant defenses of mycorrhizal fungus infection against SO_2-induced oxidative stress in Avena nuda seedlings. Bull Environ Contam Toxicol 81:440–444

Javot H, Pumplin N, Harrison MJ (2007) Phosphate in the arbuscular mycorrhizal symbiosis: transport properties and regulatory roles. Plant Cell Environ 30:310–322

Ji L, Tan W, Chen X (2019) Arbuscular mycorrhizal mycelial networks and glomalin-related soil protein increase soil aggregation in Calcaric Regosol under well-watered and drought stress conditions. Soil Tillage Res 185:1–8

Jin H, Pfeffer PE, Douds DD, Piotrowski E, Lammers PJ, Shachar-Hill Y (2005) The uptake, metabolism, transport and transfer of nitrogen in an arbuscular mycorrhizal symbiosis. New Phytol 168:687–696

Karandashov V, Bucher M (2005) Symbiotic phosphate transport in arbuscular mycorrhizas. Trends Plant Sci 10:22–29

Kaya C, Ashraf M, Sonmez O, Aydemir S, Tuna AL, Cullu MA (2009) The influence of arbuscular mycorrhizal colonisation on key growth parameters and fruit yield of pepper plants grown at high salinity. Sci Hortic (Amsterdam) 121:1–6

Khalid M, Saeed U-R, Hassani D, Hayat K, Pei Z, Nan HUI (2021) Advances in fungal-assisted phytoremediation of heavy metals: a review. Pedosphere 31:475–495

Khalvati MA, Hu Y, Mozafar A, Schmidhalter U (2005) Quantification of water uptake by arbuscular mycorrhizal hyphae and its significance for leaf growth, water relations, and gas exchange of barley subjected to drought stress. Plant Biol 7:706–712

Kistner C, Parniske M (2002) Evolution of signal transduction in intracellular symbiosis

Kong L, Gong X, Zhang X, Zhang W, Sun J, Chen B (2020) Effects of arbuscular mycorrhizal fungi on photosynthesis, ion balance of tomato plants under saline-alkali soil condition. J Plant Nutr 43:682–698

Koske RE (1981) Gigaspora gigantea: observations on spore germination of a VA-mycorrhizal fungus. Mycologia 73:288–300

Kosuta S, Chabaud M, Lougnon G, Gough C, Dénarié J, Barker DG, Bécard G (2003) A diffusible factor from arbuscular mycorrhizal fungi induces symbiosis-specific MtENOD11 expression in roots of *Medicago truncatula.* Plant Physiol 131:952–962

Krishnamoorthy R, Premalatha N, Karthik M, Anandham R, Senthilkumar M, Gopal NO, Selvakumar G, Sa T (2017) Molecular markers for the identification and diversity analysis of arbuscular mycorrhizal fungi (AMF). In: Molecular markers in mycology. Springer, Heidelberg, pp 177–199

Krzyzaniak Y, Magnin-Robert M, Randoux B, Fontaine J, Sahraoui ALH (2021) Combined use of beneficial bacteria and arbuscular mycorrhizal fungi for the biocontrol of plant cryptogamic diseases: evidence, methodology, and limits. Symbiotic Soil Microorg 429–468

Kuhn H, Küster H, Requena N (2010) Membrane steroid-binding protein 1 induced by a diffusible fungal signal is critical for mycorrhization in *Medicago truncatula.* New Phytol 185:716–733

Kumar S, Saxena S (2019) Arbuscular mycorrhizal fungi (AMF) from heavy metal- contaminated soils: molecular approach and application in phytoremediation. In: Biofertilizers for sustainable agriculture and environment. Springer, Heidelberg, pp 489–500

Latef AAHA, Chaoxing H (2011a) Arbuscular mycorrhizal influence on growth, photosynthetic pigments, osmotic adjustment and oxidative stress in tomato plants subjected to low temperature stress. Acta Physiol Plant 33:1217–1225

Latef AAHA, Chaoxing H (2011b) Effect of arbuscular mycorrhizal fungi on growth, mineral nutrition, antioxidant enzymes activity and fruit yield of tomato grown under salinity stress. Sci Hortic (Amsterdam) 127:228–233

Lehmann A, Rillig MC (2015) Arbuscular mycorrhizal contribution to copper, manganese and iron nutrient concentrations in crops–a meta-analysis. Soil Biol Biochem 81:147–158

Lehto T, Zwiazek JJ (2011) Ectomycorrhizasand water relations of trees: a review. Mycorrhiza 21:71–90. https://doi.org/10.1007/s00572-010-0348-9

Lenoir I, Fontaine J, Tisserant B, Laruelle F, Sahraoui AL-H (2017) Beneficial contribution of the arbuscular mycorrhizal fungus, *Rhizophagus irregularis*, in the protection of *Medicago truncatula* roots against benzo [a] pyrene toxicity. Mycorrhiza 27:465–476

Liu A, Hamel C, Hamilton RI, Ma BL, Smith DL (2000) Acquisition of Cu, Zn, Mn and Fe by mycorrhizal maize (*Zea mays* L.) grown in soil at different P and micronutrient levels. Mycorrhiza 9:331–336

Liu H, Chen W, Wu M, Wu R, Zhou Y, Gao Y, Ren A (2017) Arbuscular mycorrhizal fungus inoculation reduces the drought-resistance advantage of endophyte-infected versus endophyte-free *Leymus chinensis.* Mycorrhiza 27:791–799. https://doi.org/10.1007/s00572-017-0794-8

Logi C, Sbrana C, Giovannetti M (1998) Cellular events involved in survival of individual arbuscular mycorrhizal symbionts growing in the absence of the host. Appl Environ Microbiol 64:3473–3479

López-Pedrosa A, González-Guerrero M, Valderas A, Azcón-Aguilar C, Ferrol N (2006) GintAMT1 encodes a functional high-affinity ammonium transporter that is expressed in the extraradical mycelium of Glomus intraradices. Fungal Genetics Biol 43(2). https://doi.org/10.1016/j.fgb.2005.10.005

Maathuis FJM (2009) Physiological functions of mineral macronutrients. Curr Opin Plant Biol 12:250–258

Machi S (2014) FNCA Biofertilizer Manural. https://www.fnca.mext.go.jp/english/bf/bfm/e_bfm.html

Marinho F, Silva GA, Ferreira AC, Veras JDN, Sousa NMF, Goto BT, Oehl F et al (2014) Bulbospora minima, a new genus and a new species in the Glomeromycetes from semi-arid Northeast Brazil. Sydowia 66(2):313–323

Marschner H (1995) Mineral nutrition of higher plants, 2nd edn. Academic Press, pp 99–101

Mathur S, Tomar RS, Jajoo A (2019) Arbuscular Mycorrhizal Fungi (AMF) protects photosynthetic apparatus of wheat under drought stress. Photosynth Res 139:227–238

Medina A (2010) Effectiveness of the application of arbuscular mycorrhiza fungi and organic amendments to improve soil quality and plant performance under stress conditions. J Soil Sci Plant Nutrition 10:354–372

Minchin RF, Ridgway HJ, Condron L, Jones EE (2012) Influence of Inoculation with a *tricho-derma* bio-inoculant on ectomycorrhizal colonisation of *pinus radiata* seedlings. Ann Appl Biol 161:57–67. https://doi.org/10.1111/j.1744-7348.2012.00552.x

Miransari M (2011) Arbuscular mycorrhizal fungi and nitrogen uptake. Arch Microbiol

Morin C, Samson J, Dessureault M (1999) Protection of black spruce seedlings against Cylindro-cladium root rot with ectomycorrhizal fungi. Can J Bot 77:169–174

Oehl F, Sieverding E, Palenzuela J, Ineichen K, da Silva GA (2011) Advances in Glomeromycota taxonomy and classification. IMA Fungus 2:191–199

Ohyama T (2010) Nitrogen as a major essential element of plants. Nitrogen Assim Plants 37:1–17

Oliveira RS, Castro PML, Dodd JC, Vosátka M (2006) Different native arbuscular mycorrhizal fungi influence the coexistence of two plant species in a highly alkaline

Pagano MC (2014) Drought stress and mycorrhizal plant. In: Use of microbes for the alleviation of soil stresses, vol 1. Springer, Heidelberg, pp 97–110

Pagter M, Bragato C, Brix H (2005) Tolerance and physiological responses of Phragmites australis to water deficit. Aquat Bot 81:285–299

Parniske M (2008) Arbuscular mycorrhiza: the mother of plant root endosymbioses. Nat Rev Microbiol 6(10):763–775. https://doi.org/10.1038/nrmicro1987

Pérez-Tienda J, Testillano PS, Balestrini R, Fiorilli V, Azcón-Aguilar C, Ferrol N (2011) GintAMT2, a new member of the ammonium transporter family in the arbuscular mycorrhizal fungus Glomus intraradices. Fungal Genet Biol 48:1044–1055

Porcel R, Aroca R, Ruiz-Lozano JM (2012) Salinity stress alleviation using arbuscular mycorrhizal fungi. A review. Agron Sustain Dev. https://doi.org/10.1007/s13593-011-0029-x

Porcel R, Ruiz-Lozano JM (2004) Arbuscular mycorrhizal influence on leaf water potential, solute accumulation, and oxidative stress in soybean plants subjected to drought stress. J Exp Bot 55(403):1743–1750

Pozo MJ, Jung SC, López-ráez J.A (2010) Impact of Arbuscular Mycorrhizal Symbiosis on Plant Response to Biotic Stress : The Role of Plant Defence Mechanisms 193207

Püschel D, Bitterlich M, Rydlová J, Jansa J (2020) Facilitation of plant water uptake by an arbuscular mycorrhizal fungus: a Gordian knot of roots and hyphae. Mycorrhiza 30:219–330

Qiangsheng WU, Renxue XIA (2006) Effects of arbuscular mycorrhizal fungi on leaf solutes and root absorption areas of trifoliate orange seedlings under water stress conditions, pp 312–317. https://doi.org/10.1007/s11461-006-0035-3

Qin W, Yan H, Zou B, Guo R, Ci D, Tang Z, Zou X, Zhang X, Yu X, Wang Y (2021) Arbus-cular mycorrhizal fungi alleviate salinity stress in peanut: evidence from pot-grown and field experiments. Food Energy Secur e314

Ramos AC, Façanha AR, Palma LM, Okorokov LA, Cruz Z, Silva AG, Siqueira AF, Bertolazi AA, Canton GC, Melo J (2011) An outlook on ion signaling and ionome of mycorrhizal symbiosis. Brazilian J Plant Physiol 23:79–89

Raziuddin F, Hassan G, Akmal M, Salim Shah S, Mohammad F, Shafi M, Bakht J, Zhou W (2011) Effects of cadmium and salinity on growth and photosynthesis parameters of Brassica species. Pakistan J Bot 43:333–340

Read DJ (1991) Mycorrhizas in ecosystems. Experientia 47:376–391

Redecker D, Kodner R, Graham LE (2000) Glomalean fungi from the Ordovician. Science 289(5486):1920–1921. https://doi.org/10.1126/science.289.5486.1920

Redecker D, Raab P (2006) Phylogeny of the Glomeromycota (Arbuscular mycorrhizal fungi): recent developments and new gene markers. Mycologia 98:885–895. https://doi.org/10.1080/15572536.2006.11832618

Richardson AE, Barea JM, McNeill AM, Prigent-Combaret C (2009) Acquisition of phosphorus and nitrogen in the rhizosphere and plant growth promotion by microorganisms. Plant Soil 321:305–339. https://doi.org/10.1007/s11104-009-9895-2

Rillig MC, Mummey DL (2006) Mycorrhizas and soil structure. New Phytol 171:41–53. https://doi.org/10.1111/j.1469-8137.2006.01750.x

Ruissen T (2013) Arbuscular mycorrhizal fungi and their ecological roles: a review with a Norwegian perspective. Agarica 33:105–116

Ruiz-Sánchez M, Aroca R, Muñoz Y, Polón R, Ruiz-Lozano JM (2010) The arbuscular mycorrhizal symbiosis enhances the photosynthetic efficiency and the antioxidative response of rice plants subjected to drought stress. J Plant Physiol 167:862–869

Sanders IR, Clapp JP, Wiemken A (1996) The genetic diversity of arbuscular mycorrhizal fungi in natural ecosystems—a key to understanding the ecology and functioning of the mycorrhizal symbiosis. New Phytol 133:123–134

Schüßler A, Schwarzott D, Walker C (2001) A new fungal phylum, the Glomeromycota: phylogeny and evolution. Mycol Res 105:1413–1421. https://doi.org/10.1017/S0953756201005196

Selosse MA, Richard F, He X, Simard SW (2006) Mycorrhizal networks: des liaisons dangereuses? Trends Ecol Evol 21:621–628. https://doi.org/10.1016/j.tree.2006.07.003

Shibata R, Yano K (2003) Phosphorus acquisition from non-labile sources in peanut and pigeon pea with mycorrhizal interaction. Appl Soil Ecol 24:133–141

Smith SE, Smith FA (2011) Roles of arbuscular mycorrhizas in plant nutrition and growth: new paradigms from cellular to ecosystem scales. Annu Rev Plant Biol 62:227–250

Stürmer SL (2012) A history of the taxonomy and systematics of arbuscular mycorrhizal fungi belonging to the phylum Glomeromycota. Mycorrhiza 22, 247–258. Sustain Agric 2:907–930

Sylvia DM (1986) Spatial and temporal distribution of vesicular-arbuscular mycorrhizal fungi associated with Uniola paniculata in Florida foredunes. Mycologia 78:728–734

Symanczik S, Krützmann J, Nehls U, Boller T, Courty P-E (2020) Expression of major intrinsic protein genes in Sorghum bicolor roots under water deficit depends on arbuscular mycorrhizal fungal species. Soil Biol Biochem 140:107643

Van der Heijden MGA, Boller T, Wiemken A, Sanders IR (1998) Different arbuscular mycorrhizal fungal species are potential determinants of plant community structure. Ecology 79:2082–2091

Vance CP, Uhde-Stone C, Allan DL (2003) Phosphorus acquisition and use: critical adaptations by plants for securing a nonrenewable resource. New Phytol 157:423–447

Wahid F, Sharif M, Fahad S, Adnan M, Khan IA, Aksoy E, Ali A, Sultan T, Alam M, Saeed M, Ullah H, Basir A, Noor M, Khan NA (2019) Arbuscular mycorrhizal fungi improve the growth and phosphorus uptake of mung bean plants fertilized with composted rock phosphate fed dung in alkaline soil environment. J Plant Nutr 42:1760–1769. https://doi.org/10.1080/01904167.2019.1643371

Wiemken V (2007) Trehalose synthesis in ectomycorrhizas—a driving force of carbon gain for fungi?. New Phytologist 228–230

Willis A, Rodrigues BF, Harris PJC (2013) The ecology of arbuscular mycorrhizal fungi. Crit Rev Plant Sci 32(1):1–20. https://doi.org/10.1080/07352689.2012.683375

Xu H, Lu Y, Tong S (2018) Effects of arbuscular mycorrhizal fungi on photosynthesis and chlorophyll fluorescence of maize seedlings under salt stress. Emirates J Food Agric 30:199–204. https://doi.org/10.9755/ejfa.2018.v30.i3.1642

Yadav BK, Tarafdar JC (2010) Studies on phosphatase activity and clusterbean production as influenced by the P mobilizing organism Emericella rugulosa. Legum Res Int J 33:114–118

Ye L, Zhao X, Bao E, Cao K, Zou Z (2019) Effects of arbuscular mycorrhizal fungi on watermelon growth, elemental uptake, antioxidant, and photosystem II activities and stress-response gene expressions under salinity-alkalinity stresses. Front Plant Sci 10:863

Zarik L, Meddich A, Hijri M, Hafidi M, Ouhammou A, Ouahmane L, Duponnois R, Boumezzough A (2016) Utilisation de champignons mycorhiziens arbusculaires en vue de l'amélioration de la tolérance à la sécheresse de Cupressus atlantica G. Comptes Rendus - Biol 339:185–196. https://doi.org/10.1016/j.crvi.2016.04.009

Zhu X, Cao Q, Sun L, Yang X, Yang W, Zhang H (2018) Stomatal conductance and morphology of arbuscular mycorrhizal wheat plants response to elevated CO_2 and NaCl stress. Front Plant Sci 9:1363

Part IV
Urbanism, Green Building and Land Use: What can We Do in the Face of Climate Change

Chapter 10
Life Cycle Assessment for Green Buildings: A Mini Literature Review

Redouan Assadiki, Gérard Merlin, Hervé Boileau, Catherine Buhé, and Fouzi Belmir

Abstract The environmental impacts of buildings have become a major challenger during the recent years, as they produce 30% of the world's landfill waste, utilize more than 40% of the energy produced worldwide, and emit one third of all greenhouse gas emissions. This study uses systematic literature review based on the Social Science Citation Index databases to identify the publication of the building life cycle assessment (LCA) literature produced between 2000 and 2021. The significant rise in the number of global research, the United States is in the lead. When it comes to building life cycle assessments, the Norwegian University of Science and Technology is regarded as the pioneer in the field of research. In addition, the United Kingdom has contributed over 161 publications and Switzerland has contributed 84 studies. However, over the last 20 years, the United States has contributed 449 valuable papers on the LCA of buildings. The review of 53 articles reveals that the CML2002 and eco indicator 99 methods are frequently used, along with the four indicators global warming potential, photochemical ozone creation potential, and acidification potential. However, the number of LCA studies conducted in Africa is limited. Suggestions for future research are proposed to address these gaps and expand the scope of green building research in Morocco.

Keywords Systematic literature review · Life cycle assessment · Construction industry

R. Assadiki (✉) · F. Belmir
Laboratory of Materials, Processes, Catalysis, and Environment, Higher School of Technology, Sidi Mohamed Ben Abdellah University, 30000 Fez, Morocco
e-mail: redouan.assadiki@usmba.ac.ma

G. Merlin · H. Boileau · C. Buhé
LOCIE, UMR 5271, Polytech Annecy-Chambéry, Université Savoie Mont Blanc-CNRS, 73376 Le Bourget-du-Lac, France

K. Kahime et al. (eds.), *Climate Change Effects and Sustainability Needs*, Springer Climate, https://doi.org/10.1007/978-3-031-59603-2_10

Introduction

Morocco's construction sector is characterized by rapid development an average of 170,000 builds per year (United Nations Environment 2016). To address the estimated demand for new housing units, the Moroccan government has implemented an ambitious building program. This sector, which accounts for nearly one-third of national energy consumption, has a vital contribution to make to Morocco's economic and social development, contributing 7% of national GDP (United Nations Environment 2016). Over 25% of Morocco's primary energy consumption comes from the building sector (residential and tertiary), and that percentage could rise to 40% by 2030 if building life cycles are taken into account. In particular, the construction industry accounts for 40–50% of the world's greenhouse gas emissions due to its consumption of 40% of materials that enter the global economy (Pachauri et al. 2014). By quantifying environmental impacts and presenting workable scenarios for mitigating environmental impacts, Life Cycle Assessment (LCA) offers an appropriate method for accomplishing this goal. Certainly, LCA has been a mature and widely used method for the assessment of environmental impacts. The GHG Protocol, ISO 14067, and Publicly Available Specification 2050 (PAS2050) are the three most widely used standards for conducting an LCA of products or services. Although the computation and exclusion criteria for these standards vary, they are all based on ISO 14040 and ISO 14044, which offer the guiding principles, essential requirements, and structure for life cycle evaluations. LCA is utilized through various sustainability certification systems such as BREEAM, LEED, DGNB, and HQE (Sartori et al. 2021).

Reviews of the literature on building life cycle assessment have been published; these reviews tend to concentrate on bibliometrics, and the systematic literature covers only a small portion of the macro-scale research topics and trends. Chau et al. (2015) reviewed on three streams of life cycle studies LCA, Life Cycle Energy Assessment, and Life Cycle Carbon Emissions Assessment. Geng et al. (2017) and Nwodo and Anumba (2019) analyzed the literature on building life cycle assessment (LCA) published on databases of web of science. Minunno et al. (2021) identified the common topics of LCA, embodied energy, and embodied carbon of construction material through the systematic literature reviews such as concrete, cement, steel, and wood. Rhaouti et al. (2023) compared the results of 32 LCA studies of the 11 cement types according to the GWP indicator. Analyses of the effects of thermal systems, building integrated renewable energy systems, and heating and cooling systems on the environment have also been done in the past.

Finding out the features of the worldwide LCA literature for buildings from 2000 to 2021 is the aim of this study. The systematic literature review (SLR) is to examine global trends in building LCA research over the past 21 years by analyzing general trends in the methodologies used in the impact inventory, databases, tools, and LCA software. In the absence of similar studies in Morocco, these results provide us with useful information to select the method we can use or adapt in the Moroccan context, and to propose sub-themes in future research direction on the LCA building.

Life Cycle Assessment in the Building Sector

Life Cycle Assessment is an approach for methodically accounting for the inputs, outputs, and possible environmental consequences of a product system throughout its life cycle, often from cradle to grave. LCA's comprehensive approach can highlight possible trade-offs across life cycle phases and distinct environmental consequences, allowing for the comparison and assessment of alternative solutions. LCA has acquired universal acceptability in industry as an environmental assessment technique and began to be utilized in the construction context approximately twenty years ago to examine both building materials and structures. The ISO 14040 set of standards defines the overall framework for LCA applications (ISO 14040 2006; ISO 14044 2006; ISO 14049 2012), covering the four interconnected steps for an LCA application: aims and scope, Life Cycle Inventory (LCI), Life Cycle Impact Assessment (LCIA), and results interpretation.

The guidelines have served as a foundation for LCA applications in the construction industry. Furthermore, ISO has defined Type III environmental declarations (ISO 14025 2006), often known as environmental product declarations (EPD). LCA is not a green label, but rather a technique for quantifying and recognizing the environmental consequences of buildings throughout their life cycle, estimating the relative performance of various construction choices, and perhaps identifying how and where improvements might be made. Moreover, standardization on life cycle assessments (LCA) of buildings (EN 15978; CEN 2011) and construction products (EN 15804; CEN 2012) have been developed in the European by the technical committee on the sustainability of construction works (CEN/TC 350). This work defines specific steps or modules that can be used in LCA studies to facilitate easier comparability between studies. A life cycle approach is used to analyze the environmental performance of Green Building, and the concepts, standards, and processes for this evaluation are outlined in the European standard EN 15804:2012. It offers a framework for evaluating the environmental effects of building goods over their whole life cycle (Fig. 10.1), including raw material extraction (A1–3), transportation (A4), manufacturing (A5), usage (B1–7), and disposal (C1–4).

The EN 15804:2012 standard, designed for manufacturers of construction products, architects, engineers and other construction professionals, aims to establish a standardised methodology for assessing and reporting the environmental impacts of construction products. Its use is intended to create a common approach to the environmental assessment of products, making it easier to compare different construction products. This standard has a critical function in supporting the development of sustainable construction practices, by providing uniform criteria and guidelines for assessing the environmental performance of products, thereby promoting informed decision-making in the construction sector.

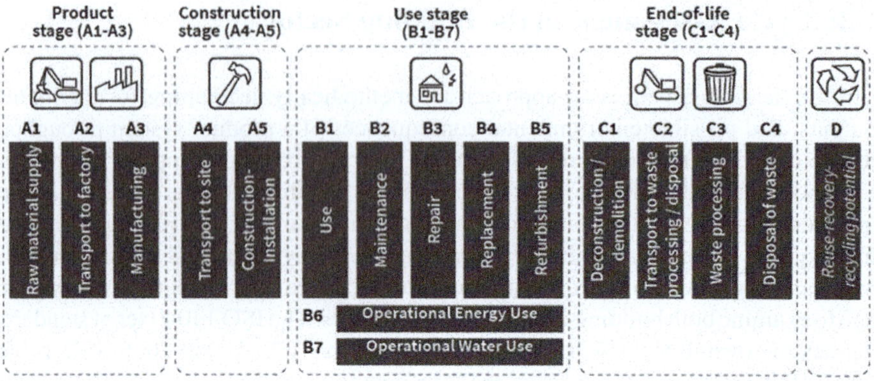

Fig. 10.1 Building assessment modules for life cycle assessment according to EN 15804:2012/EN 15978

Research Methodology

This review focused study employs a systematic literature review (SLR) to analyze journal articles and conference proceedings published between 2000 and 2021. In order to accomplish the goals of this investigation, the integrated analysis illustrated in Fig. 10.2.

First, a collection of pertinent literature on the life cycle analysis of buildings is presented, along with information on other building-related topics including "building materials," "building systems," and "green building." Additionally, "build" has been swapped out for "construction" or "architecture". Consequently, almost 2500 articles were taken from databases that contained content released between 2000 and 2021. Were based on the review of titles and summaries in databases of science-indexed scientific citations (DTU Find It, Google Scholar, Scopus, Springer,

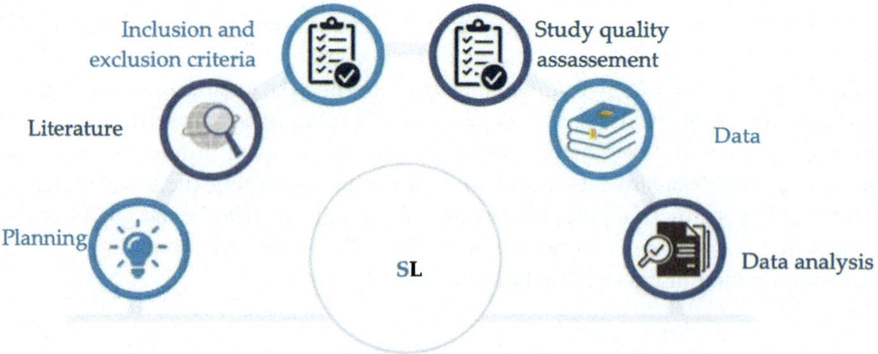

Fig. 10.2 Summary of the research design

and Web of Science). In this phase, the search criteria were applied to identify certain words or phrases in the keywords, abstract, and title of possible entries. Second, in order to remove redundant datasets, the chosen research was examined again. In order to accomplish the goals of this investigation, the integrated analysis depicted in Fig. 10.2 was carried out. It comprises of an organized study of the literature on sustainable building and sustainable building assessment techniques found in bibliographic records.

Third, the entire texts of the remaining papers were assessed in order to determine which publications should be reviewed. Project kinds that were irrelevant, knowledge domains that were not addressed, and complex evaluation methods that overlooked the life cycle were among the exclusion criteria. To achieve 53 articles on life cycle analysis, there should be a recognized focus on relevant articles related to such a sector. The literature reviewed to identify the main methodologies developed in the life cycle impact assessment which are our goal. The environmental indicators are used to express the impact of the building on the environment and identify potential gaps in futures research.

Results and Discussion

Overview of Life Cycle Studies of Buildings

According to the findings, life cycle assessment has primarily focused on topics related to engineering, environmental sciences and ecology, and construction and building technology, with the terms "building envelope" and "living wall" developing increased attention in recent years, as evidenced by citation bursts. Processes (bottom–up), input–output (IO) data (top–down), and hybrid analysis (including process and IO data) can all be used in the Life Cycle Impact Assessment. The process-based method is the most commonly utilized in building LCA studies because it gives extensive information at the product level, allowing individual construction alternatives to be compared (Monteiro et al. 2016). The base of top-down analysis is the correlation between the energy intensity data per industrial sector and the domestic economic statistics for various industry sectors. Although IO analysis yields more comprehensive results at the macroscale, it is more unpredictable due to "aggregation error," which makes it challenging to describe the effects on various processes within the same sector (Crawford 2008). In order to solve the shortcomings of process-based LCA and IO analysis, the IO-hybrid technique was created. According to the authors that have used the IO-hybrid technique, an IO-hybrid LCA can produce far higher embodied energy estimates than comparable process-based studies since it has more thorough system boundaries (Stephan 2014). Throughout the 53 publications related to LCA building as shown in Table 10.1.

Figure 10.3 presents the results drawn from 53 papers extracted from data repositories. The CML (23%) and eco-indicator 99 (15%) methods stand out as the preferred

Table 10.1 Publications that received the highest citations per year regarding building LCA 2000–2021

Author	Location	Methodology	Indicator environmental
Scheuer et al. (2003)	USA	Midpoints + Cum. En.	GWP, ODP, NP, AP, waste
Erlandsson and Levin (2004)	Sweden	BYKR	AP, EP, GWP, POCP, LUB
Xing et al. (2008)	China	Midpoints	Energy use, GWP, l pollution emissions (O_3, CO, NO_x, PM10, SO_x)
Kofoworola and Gheewala (2008)	Thailand	EIO-LCA	GWP + AP + POCP
Ortiz-Rodríguez et al. (2010)	Spain	CML 2	GWP, AP, RD, HTP, EP, CEDnr, CEDr, water
Guardigli et al. (2011)	Italy	Eco-Ind.99	Human health, ecosystem quality, and resources
Cuéllar-Franca and Azapagic (2012)	UK	CML 2001	GWP, AP, ADP, EP, ODP, HTP, TETP, FAETP, MAETP, POCP
Peuportier et al. (2013)	France	–	En, GWP, AP, EP, biodiversity, odour, resource, smog, waste, water
Dong and Ng (2014)	Hong Kong	ReCiPe + CML + eco-indicator 99	GWP, AP, EP, ODP, HTP, TETP, FAETP, MAETP, POCP
Sofia (2015)	Portugal	CML + EF + indicator 99 + carbon footprint	GWP, AP, EP, ODP, ADP, POCP
Zaini et al. (2016)	Malaysia	Em.Energy	GWP
Alba-Rodríguez et al. (2017)	Spain	EF + D.En	Pastures, sea, cropland, forest, energy, built land EF (gha)
Tumminia et al. (2018)	Italy	CML	GWP, AP, EP, ODP, HTP, GER, POCP, ADP
Emami et al. (2019)	Finland	ReCiPe	GWP, ODP, TA, WD, EP, HTP, POCP, PM, TETP, FAETP, MAETP, I, RD
Ryberg et al. (2021)	Denmark	ReCiPe	GWP, ODP, TA, WD, LUB, EP, HTP, POCP, PM, TETP, RD, FAETP, MAETP, I, TE

EF Ecological Footprint, *Eco-Ind.99* Eco-indicator 99, *CML* Institute of Environmental Science, *Cum. En.* Cumulated Energy, *EIO-LCA* an economic input-output life-cycle assessment, *BYKR* Swedish Building Eco-Cycle Council, *CEDnr* use of non-renewable primary energy, *CEDr* use of renewable primary energy (energy resources), *GER* Gross Energy Requirement, *ODP* depletion potential of the stratospheric ozone layer, *POCP* formation potential of tropospheric ozone photochemical oxidant, *TETP* Terrestrial ecotoxicity potential, *ADP* Abiotic depletion potential, *MAETP* Marine aquatic ecotoxicity potential, *TA* Terrestrial Acidification, *TE* Terrestrial Eutrophication, *WD* Water Depletion, *LUS* Land Use Soil Erosion, *LUB* Land Use Biodiversity, *HTP* Human toxicity potential, *EP* Marine Eutrophication, *RD* resource depletion, *PM* Particulate, *NP* Nitrification potential, *FAETP* Freshwater aquatic ecotoxicity potential, *I* Ionizing Radiation

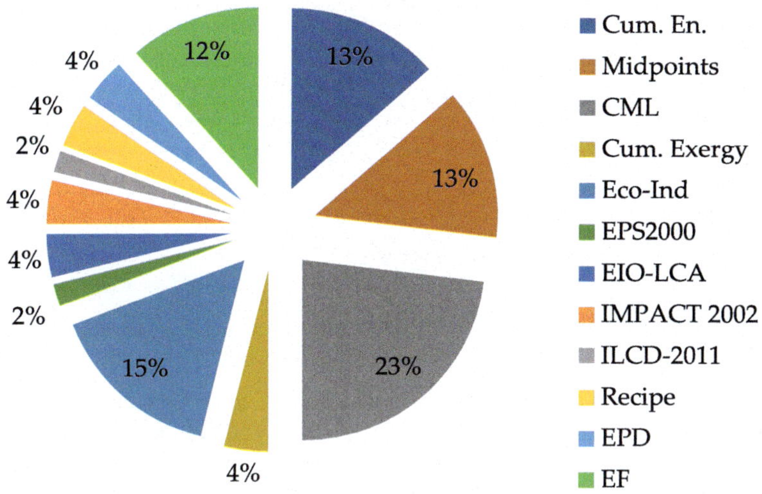

Fig. 10.3 The methodologies used in the life cycle impact assessment building (2000–2021)

approaches for assessing environmental impacts, indicating their notable popularity in the reviewed literature. Conversely, the other reviewed methods show a lower use, suggesting either a restricted preference on the part of the researchers or a restricted applicability to particular contexts, highlighting the variety of methodological options available in the field of environmental assessment.

To compare the two methods, the table summarizes the difference between the Eco-indicator and CML 2002 (Table 10.2). CML (Centre for Environmental Science at Leiden University) is a widely used impact assessment method that was developed in the Netherlands. It consists of a set of environmental impact categories, such as global warming, acidification, and eutrophication, and a set of characterization factors that quantify the potential environmental impacts of different pollutants in each category. Indicator 99 is an impact assessment method that was developed by the European Commission Joint Research Centre. It is similar to CML in that it consists of a set of environmental impact categories and characterization factors, but it also includes additional impact categories such as human toxicity and land use. The CML method is a midpoint (cause or effect) method that uses eleven environmental indicators, except for energy, whereas the methodology indicators-99 (consequence) uses indicators: Human Health (DALY), Quality of the ecosystem Resources (energy required).

Overall, these two approaches continue the shared goal of offering a thorough assessment of the environmental effects of products or processes. However, the distinctions between them lie in the specific effect categories and the defining factors they employ. The choice between the CML method and the indicators-99 methodology depends on the specific goals and scope of the analysis of the Cycle of Life (LCA) study, highlighting the significance of choosing a methodological approach based on the factors to be evaluated.

Table 10.2 The comparison between the two methods

Methods	Midpoint	Endpoint	Characterization	Normalization	Ponderation	Single score
CML 2002 (De Bruijn et al. 2002)	×	–	×	×	–	–
Eco-indicator 99 (Goedkoop 1999)	–	×	×	×	×	×

Environmental Performance Indicators

The indicators most frequently used in research carried out between 2000 and 2021 include the 100-year global warming potential, the acidification potential, the photochemical oxidant formation potential of tropospheric ozone, and the eutrophication potential. This prevalence can be explained by the fact that these indicators are specifically chosen to measure the impacts of contemporary climate change and ozone depletion. These choices reflect the increased relevance of these aspects in understanding current environmental consequences, as highlighted in Fig. 10.4.

Even though accounting for primary energy or greenhouse gas emissions is representative of some environmental categories like global warming potential (GWP), acidification potential (AP), and abiotic depletion (AD), it may not fully represent the

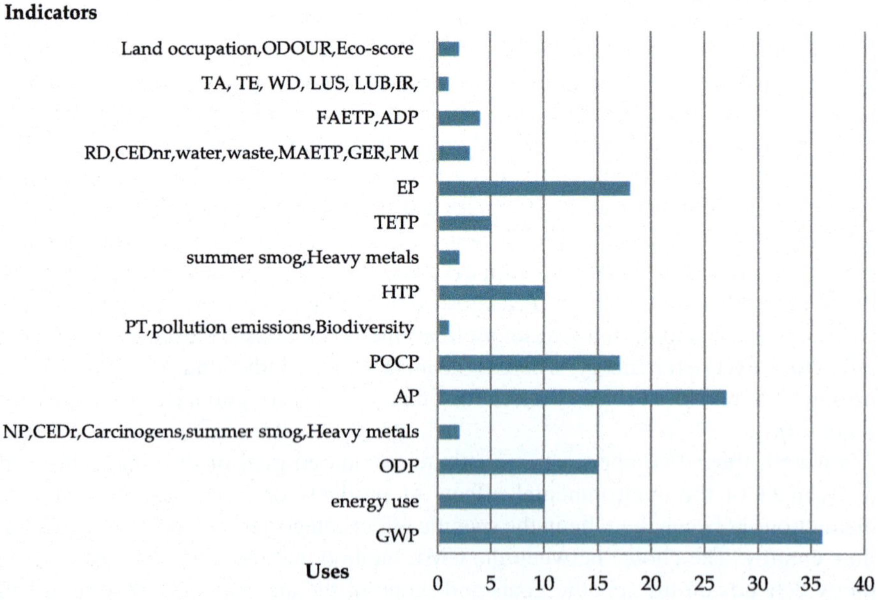

Fig. 10.4 Indicators of environmental assessment

overall environmental impact of a building, including other categories like eutrophication potential (EP), photochemical oxidation (PO), ozone layer depletion (OLD), and toxicity categories. This was highlighted by the most thorough life cycle assessment (LCA) studies that included multiple environmental categories. Since there is no required technique for conducting an environmental impact assessment, some research indicates that the environmental categories chosen and the methodology employed for conducting an environmental impact assessment may have an impact on the assessment's outcomes.

LCA Software and LCI Databases

The use of LCA software is currently at the very peak of development, with several software tools available and new ones being created on a regular basis. Table 10.3 lists software solutions suitable for general assessments, i.e. not specialised in a specific application, based on 53 articles. The LCI data relating to a product or process, together with the outputs and accompanying material, energy and emission flows, are stored in the LCA software databases. Where primary data is not available, as is often the case for upstream and downstream processes, the LCI databases are used.

The process of choosing relevant data sets requires a lot of work, even though using LCI databases minimizes the effort required for data collecting. Since the first LCI databases appeared in the 1990s, several datasets have been developed by

Table 10.3 Overview of LCA tools and LCI databases (2000–2022)

Software/tool	Sector	Database	Number of datasets	Sector	Free/for purchase
SimaPro	Generic	Ecoinvent (worldwide)	18,000+	Generic	3800$/ multi-user
GaBi	Generic	GaBi (worldwide)	15,500+	Generic	1500$/user
OpenLCA	Generic	ProBas (worldwide)	9000+	Generic	Free
Umberto 11	Generic	IDEMAT (worldwide)	1700+	Generic	Free (academic)
LCA Calculator	Generic	Ökobaudat (Germany)	1250+	Construction	Free
Athena Software	Building material	MTU asphalt pavement framework (US)	7000+	Construction	Free
LCA Calculator	Generic	MTU asphalt (US)	8500+	Construction	Free
		PSILCA (worldwide)	10,000+	Generic (social LCA)	3000$/user

various suppliers across various industries and geographical areas; some are available for free, while others require payment. Choosing relevant datasets and keeping an overview is challenging due to this variability. Therefore, Table 10.3 gives a summary of input–output databases, social LCA databases, and common LCI databases. It is not claimed that the compilation is exhaustive. It is to be expected in emerging nations like Morocco.

In Morocco, life cycle assessment (LCA) has been applied in two key sectors: agricultural products for export and photovoltaic (PV) installations. An in-depth study was carried out, particularly in the agricultural sector, where An LCA study was conducted, in which the environmental impact of one kilogram of tomatoes was assessed. The results show that 0.306 kg CO_{2eq} is emitted during the production and 0.204 kg of CO_{2eq} is emitted during transport to France. These are significantly detailed figures for the carbon footprint associated with the production and distribution of this specific product. In addition, the study took into account the water footprint, revealing a total consumption of 29.738 L of water per kilogram of tomatoes (Payen et al. 2015). This level of detail in the analysis provides an in-depth understanding of the environmental impacts associated with the production and export of agricultural products. As well, the Corona et al. (2017) study was carried out in 2016, focusing on the NOOR project in Ouarzazate, which uses high-concentration photovoltaic technology. The analysis quantified the environmental aspects of large-scale solar energy production, providing crucial information on greenhouse gas emissions, water consumption and other impacts associated with this type of installation.

During the two Life Cycle Assessment (LCA) studies, the method used has been to use Ecoinvent's generic databases. Although this approach is very common, it raises questions about the robustness of the results obtained for the environmental indicators. The generic nature of these databases may introduce a degree of uncertainty into environmental assessments specific to the Moroccan context. It is crucial to address this issue of uncertainty, as it can significantly influence the interpretation of the results and the relevance of the resulting recommendations.

Criticisms Relative to Life Cycle Analysis

"LCA is one of the most reliable tools known to date for the entire industrial production system" (Côté and Gouvinhas 2004). Also, this method is considered the most valid in terms of evaluation. The analysis of scientific publications related to Life Cycle Assessments of complex products revealed barriers associated with carrying out such assessments. These barriers were used to develop search strings for the main literature search, which aimed to identify a broad range of approaches, not just those declared as simplification approaches. The most frequently cited barrier to carrying out LCA of complex products was the time and resource requirements, primarily due to the data-intensive life cycle inventory phase. Other challenges included the complexity of the product, data and resource constraints, and required expert knowledge. Since buildings and other goods contain a wide variety of parts and materials, it is very important to put in a lot of work to compile and organize data. Modelling all parts and materials is often not feasible, and the use of generic databases of components purchased from suppliers may not contain the required data. In addition, industrial building considers the LCA methodology to be too complex to be used routinely. The uncertain results come because of the absence of the spatial or temporal dimension in the data. In addition, the life cycle analysis does not take into account social, economic, technical, strategic or regulatory aspects.

Notwithstanding the weaknesses of LCA and the completeness of the inventory stage (database), this is still a hot topic in Morocco. Given the absence of databases on construction materials and products (Environmental Product Declarations), we propose some suggestions to launch the research:

i. Implementing Environmental Product Declarations Morocco.
ii. Simplification of LCA based on detailed and validated models.
iii. Integration with other assessment methods to simplify and ensure consistency of the data used.
iv. Integrate machine learning with Life Cycle Assessment (LCA), enabling LCA data to be used optimally and thus boosting the accuracy and efficiency of environmental assessments for more informed and sustainable decision-making.

Conclusions

The building sector is characterised by its large diversity of materials, construction processes and long service life (50–100 years). There will always be new advancements in building materials and techniques, which will cause the gap in the present databases' sufficiency to grow and increase the need for more study. Accurate comparisons across study findings for future references will be hampered by the varied LCA research technique with its own aims, scopes, and constraints. The literature review shows an abounding use of the CML2002 and eco-indicator 99 methods, as well as the four indicators GWP, AP, POCP, and EP. In a constantly evolving Moroccan context, characterised by rapid advances in construction practices and growing demand for sustainable buildings, may accentuate the need to update and adapt the databases used in life cycle assessments (LCAs) to accurately reflect local realities. This underlines the relevance of more in-depth studies and standardisation of the LCA technique for buildings, not only to fill current gaps but also to anticipate developments specific to the Moroccan context.

Integrating these considerations, the results of LCA studies can have a direct impact by providing crucial information to political decision-makers, designers, engineers and building users in Morocco. They could influence the choice of sustainable construction materials and techniques, thereby contributing to the development of sustainable development in the Moroccan construction sector. Despite the current challenges of LCA, this approach remains relevant and necessary in Morocco to guide decisions towards more ecological and socially responsible construction.

Notwithstanding the weaknesses of LCA and the completeness of the inventory stage (database), this is still a hot topic in Morocco.

Finally, this mini literature review identified the most effective institutions for establishing LCA research, which may also encourage future international research collaboration with these universities.

References

Alba-Rodríguez MD, Martínez-Rocamora A, González-Vallejo P, Ferreira-Sánchez A, Marrero M (2017) Building rehabilitation versus demolition and new construction: economic and environmental assessment. Environ Impact Assess Rev 66:115–126. https://doi.org/10.1016/j.eiar.2017.06.002

Chau CK, Leung T, Ng W (2015) A review on life cycle assessment, life cycle energy assessment and life cycle carbon emissions assessment on buildings. Appl Energy 143:395–413. https://doi.org/10.1016/j.apenergy.2015.01.023

Corona B, Escudero L, Quéméré G, Luque-Heredia I, San Miguel G (2017) Energy and environmental life cycle assessment of a high concentration photovoltaic power plant in Morocco. Int J Life Cycle Assess 22(3):364–373. https://doi.org/10.1007/s11367-016-1157-y

Côté C, Gouvinhas RP (2004) Evaluation comparative de deux méthodes d'analyse de cycle de vie simplifiée dans un contexte de conception de produit au sein de PME. In: Proceedings of the Canadian engineering education association (CEEA)

Crawford RH (2008) Validation of a hybrid life-cycle inventory analysis method. J Environ Manage 88(3):496–506. https://doi.org/10.1016/j.jenvman.2007.03.024

Cuéllar-Franca RM, Azapagic A (2012) Environmental impacts of the UK residential sector: life cycle assessment of houses. Build Environ 54:86–99. https://doi.org/10.1016/j.buildenv.2012.02.005

De Bruijn H, Van Duin R, Huijbregts MAJ, Guinee JB, Gorree M, Heijungs R, Huppes G, Kleijn R, De Koning A, Van Oers L, Wegener Sleeswijk A, Suh S, Udo De Haes HA (eds) (2002) Handbook on life cycle assessment: operational guide to the ISO standards, vol 7. Springer Netherlands. https://doi.org/10.1007/0-306-48055-7

Dong YH, Ng ST (2014) Comparing the midpoint and endpoint approaches based on ReCiPe—a study of commercial buildings in Hong Kong. Int J Life Cycle Assess 19(7):1409–1423. https://doi.org/10.1007/s11367-014-0743-0

Emami N, Heinonen J, Junnonen J, Laine J, Junnila S (2019) A life cycle assessment of two residential buildings using two different LCA database-software combinations: recognizing uniformities and inconsistencies. Buildings 1–20. https://doi.org/10.3390/buildings9010020

Erlandsson M, Levin P (2004) Environmental assessment of rebuilding and possible performance improvements effect on a anational scale. Build Environ 39(12):1453–1465. https://doi.org/10.1016/j.buildenv.2004.06.001

Geng S, Wang Y, Zuo J, Zhou Z, Du H, Mao G (2017) Building life cycle assessment research: a review by bibliometric analysis. Renew Sustain Energy Rev 76:176–184. https://doi.org/10.1016/j.rser.2017.03.068

Goedkoop MJ (1999) The eco-indicator 99: a damage oriented method for life cycle impact assessment methodology report. Pre Consultants

Guardigli L, Monari F, Bragadin MA (2011) Assessing environmental impact of green buildings through LCA methods: a comparison between reinforced concrete and wood structures in the European context. Procedia Eng 21:1199–1206. https://doi.org/10.1016/j.proeng.2011.11.2131

Kofoworola OF, Gheewala SH (2008) Environmental life cycle assessment of a commercial office building in Thailand. Int J Life Cycle Assess 498–511. https://doi.org/10.1007/s11367-008-0012-1

Minunno R, O'Grady T, Morrison GM, Gruner RL (2021) Investigating the embodied energy and carbon of buildings: a systematic literature review and meta-analysis of life cycle assessments. Renew Sustain Energy Rev 143:110935. https://doi.org/10.1016/j.rser.2021.110935

Monteiro H, Fernández JE, Freire F (2016) Comparative life-cycle energy analysis of a new and an existing house: the significance of occupant's habits, building systems and embodied energy. Sustain Cities Soc 26:507–518. https://doi.org/10.1016/j.scs.2016.06.002

Nwodo MN, Anumba CJ (2019) A review of life cycle assessment of buildings using a systematic approach. Build Environ 162:106290. https://doi.org/10.1016/j.buildenv.2019.106290

Ortiz-Rodríguez O, Castells F, Sonnemann G (2010) Life cycle assessment of two dwellings: one in Spain, a developed country, and one in Colombia, a country under development. Sci Total Environ 408(12):2435–2443. https://doi.org/10.1016/j.scitotenv.2010.02.021

Pachauri RK, Allen MR, Barros VR, Broome J, Cramer W, Christ R, Church JA, Clarke L, Dahe Q, Dasgupta P (2014) Climate change 2014: synthesis report. In: Contribution of working groups I, II and III to the fifth assessment report of the Intergovernmental Panel on Climate Change. IPCC

Payen S, Basset-Mens C, Perret S (2015) LCA of local and imported tomato: an energy and water trade-off. J Clean Prod 87:139–148. https://doi.org/10.1016/j.jclepro.2014.10.007

Peuportier B, Thiers S, Guiavarch A (2013) Eco-design of buildings using thermal simulation and life cycle assessment. J Clean Prod 39:73–78. https://doi.org/10.1016/j.jclepro.2012.08.041

Rhaouti Y, Taha Y, Benzaazoua M (2023) Assessment of the environmental performance of blended cements from a life cycle perspective: a systematic review. Sustain Prod Consum 36:32–48. https://doi.org/10.1016/j.spc.2022.12.010

Ryberg MW, Ohms PK, Møller E, Lading T (2021) Comparative life cycle assessment of four buildings in Greenland. Build Environ 204. https://doi.org/10.1016/j.buildenv.2021.108130

Sartori T, Drogemuller R, Omrani S, Lamari F (2021) A schematic framework for life cycle assessment (LCA) and green building rating system (GBRS). J Build Eng 38:102180. https://doi.org/10.1016/j.jobe.2021.102180

Scheuer C, Keoleian GA, Reppe P (2003) Life cycle energy and environmental performance of a new university building: modeling challenges and design implications. Energy Build 35(10):1049–1064. https://doi.org/10.1016/S0378-7788(03)00066-5

Sofia D (2015) Environmental footprint of modular building solution using LCA approach, pp 1–15

Stephan L (2014) Reducing the total life cycle energy demand of recent residential buildings in Lebanon. https://doi.org/10.1016/j.energy.2014.07.028

Tumminia G, Guarino F, Longo S, Ferraro M, Cellura M, Antonucci V (2018) Life cycle energy performances and environmental impacts of a prefabricated building module. Renew Sustain Energy Rev 92:272–283. https://doi.org/10.1016/j.rser.2018.04.059

United Nations Environment (2016) Plan Sectoriel "Eco-Construction et Bâtiment Durable" Maroc 2016—Plan Cadre National pour les MCPD. SwitchMed Programme. https://wedocs.unep.org/xmlui/handle/20.500.11822/34046

Xing S, Xu Z, Jun G (2008) Inventory analysis of LCA on steel- and concrete-construction office buildings. Energy Build 40(7):1188–1193. https://doi.org/10.1016/j.enbuild.2007.10.016

Zaini N, Ibrahim SH, Baharun A, Nawi MNM (2016) A review of embodied energy (EM) analysis of industrialised building system (IBS). J Mater Environ Sci 7(4):1357–1365

Chapter 11
The Impact of Industrial Activities on the Land Surface Temperature (LST) in the Urban Environment of the City of Kenitra (Morocco)

Rhita Nasraoui and Khalid Benzidiya

Abstract This work presents a diagnosis of the impact of industrial activities on the land surface temperature (LST) in the industrial area of the city of Kenitra. This study is based on the determination of the variables of surface temperature and humidity within the industrial area of the city of Kenitra based on statistical and analytical methods. The aim of this study is to develop solutions to avoid negative impacts of LST on the economy and the environment. Moreover, most recent studies deal mainly with only a few of the most significant impacts, and thus future research should be redirected towards neglected indicators and relationships, which are fundamental to design effective climate policies in industrial destinations. In this paper, remote sensing technology was used to analyze changes in LST and spectral indices.

Keywords Industrial activities · Surface temperature · Kenitra · Remote sensing

Introduction

The surface temperature of the earth is important in climate studies due to its impact on the lower atmosphere's thermal equilibrium (Dibs and Alnajjar 2013). The city surface and atmosphere are exposed to urban climate, being warmer than non-urban environments, especially at night. Many factors affect the radiant energy that reaches the surface of the earth, including atmospheric transparency (Huang et al. 2021), daylight length, seasons, and radiation angle.

R. Nasraoui (✉) · K. Benzidiya
Laboratory Dynamics of Landscapes, Risks and Heritage, Faculty of Arts and Humanties, Sultan Moulay Slimane University, Beni Mellal, Morocco
e-mail: rhita.nasraoui@gmail.com

K. Benzidiya
Laboratory Dynamics of Landscapes, Risks and Heritage, Polydisciplinary Faculty of Khouribga, Sultan Moulay Slimane University, Beni Mellal, Morocco

K. Kahime et al. (eds.), *Climate Change Effects and Sustainability Needs*, Springer Climate, https://doi.org/10.1007/978-3-031-59603-2_11

Alterations in land use and surface crust result in modifications of the physical properties of the Earth's surface, which have an impact on various factors that contribute to regional climates, including radiation, heat, and water vapor (Kafy et al. 2020).

The climate of a given city is a function of its ongoing activities. Additionally, one of the common problems in each city is the formation and concentration of heat islands, which increases the consumption of electricity/water, biodiversity changes, and this, however, affects human's comfort (Bahi et al. 2016).

The satellites such as ASTER, MODIS and Landsat are useful tools to model the impact of urban heat islands and to evaluate the surface temperature as a function of the materials used in the city. Urban and semi-urban areas, whether in developed or developing countries, often suffer from poor air quality and heat islands caused by industrial activities, vehicle use, coal burning, agricultural activities, and unplanned land use.

The study underscores the importance of utilizing infrared thermal data to comprehend changes in surface temperatures in the Casablanca region of Morocco. Remote sensing data, including Landsat TM/ETM+/OLI-TIRS images from 1984 to 2016, and MODIS nighttime images acquired from 2005 to 2015, were employed to evaluate the intensity of urban surface nighttime heat islands. Urban areas exhibited higher surface temperatures due to population density and the presence of impermeable urban surfaces (Bahi et al. 2016). The research (Garouani et al. 2021) also highlights the significance of considering city morphology and employing LST (Land Surface Temperature) data in urban planning and development strategies, particularly in semi-arid regions. This involves analyzing the spatial distribution of temperature and identifying local climate zones (LCZs) using high-resolution satellite imagery and statistical methods.

The study sheds light on the importance of considering the morphology of the city of Ben Guerir and LST data in urban planning and development strategies, especially in semi-arid areas, for analyzing the spatial distribution of temperature and identifying local climate zones (LCZs). It explores the relationship between surface temperature and city morphology using moderate-resolution satellite images and a statistical approach (Azmi et al. 2021).

The study conducted by Sébastien Gadal (Gadal 2008) emphasizes the importance of using infrared thermal data to study urban concentrations in Morocco and understand the impact of urbanization on human societies. This information can assist in identifying areas most affected by urban heat and evaluating the effects of urbanization on air and water quality and the local climate.

This work aims to evaluate the use of Landsat satellite data to assess differences in surface temperatures in urban areas and to analyze and compare the relationship between urban surface temperature and land cover types. Additionally, the study employs a comprehensive approach that combines remote sensing techniques, geographic information systems, and statistical analysis. This research article contributes to the field of urban climate science by shedding light on the impact of industrial activities on surface temperature and the formation of urban heat islands in the region. Furthermore, the article emphasizes the importance of remote sensing in assessing the thermal conditions within urban areas.

Field of Study

The city of Kenitra is one of the most important cities in Morocco's northwestern region. It is located on the left-bank of the Sebou's River and 12 km from the beach of Mehdia. Therefore, this forms its estuary on the Atlantic Ocean. Thus, for administrative settings, the study area belongs to the Rabat-Sale-Kénitra region. It is bordered to the north by the regions of Larache and Ouazzane and to the west by the Atlantic Ocean.

To the east, it is adjacent to the regions of Sidi Kacem and Sidi Slimane, while the Sale's region and the Khemisset's prefecture form its southern border. The after-mentioned city is composed of 5 circles, 3 urban municipalities, and 20 rural municipalities.

The city's geographic location, which is located at the intersection of roads connecting northern and eastern cities to the south, has contributed to attracting a large number of immigrants from the other regions of the kingdom. Consequently, this is leading to the development of the urban area and the intensification of its infrastructure and industrial units.

This has contributed to the expansion of its urban area and the increase in the population, which has moved from 573,000 people in 2004 to 711,713 people in 2022. As a result, this is due to the convergence of migration movements from various regions, especially from rural areas.

The city of Kenitra is located at the western end of the Gharb plain, which is characterized by its weak slopes with an average elevation of about 60 m at the level of the sand dunes overlooking the ocean. This has led to environmental degradation, especially in the absence of integrated preventive plans for the social and economic development of the region (Fig. 11.1).

Materials and Methods

This study relied primarily on satellite data for the study years 1995 and 2021 using geographic information systems and remote sensing to calculate vegetation indices, derive land surface temperature, monitor land cover patterns, and land use. These satellite images were downloaded from of the United States Geological Survey's website (USGS). Table 11.1 shows the study area and the dates of the satellite images used.

The satellite imagery data from LANDSAT5 TM and LANDSAT8 OLI/TIRS were processed according to the methodology outlined in Fig. 11.2 to derive the surface temperature to detect both the urban heat island and the changes in land cover patterns

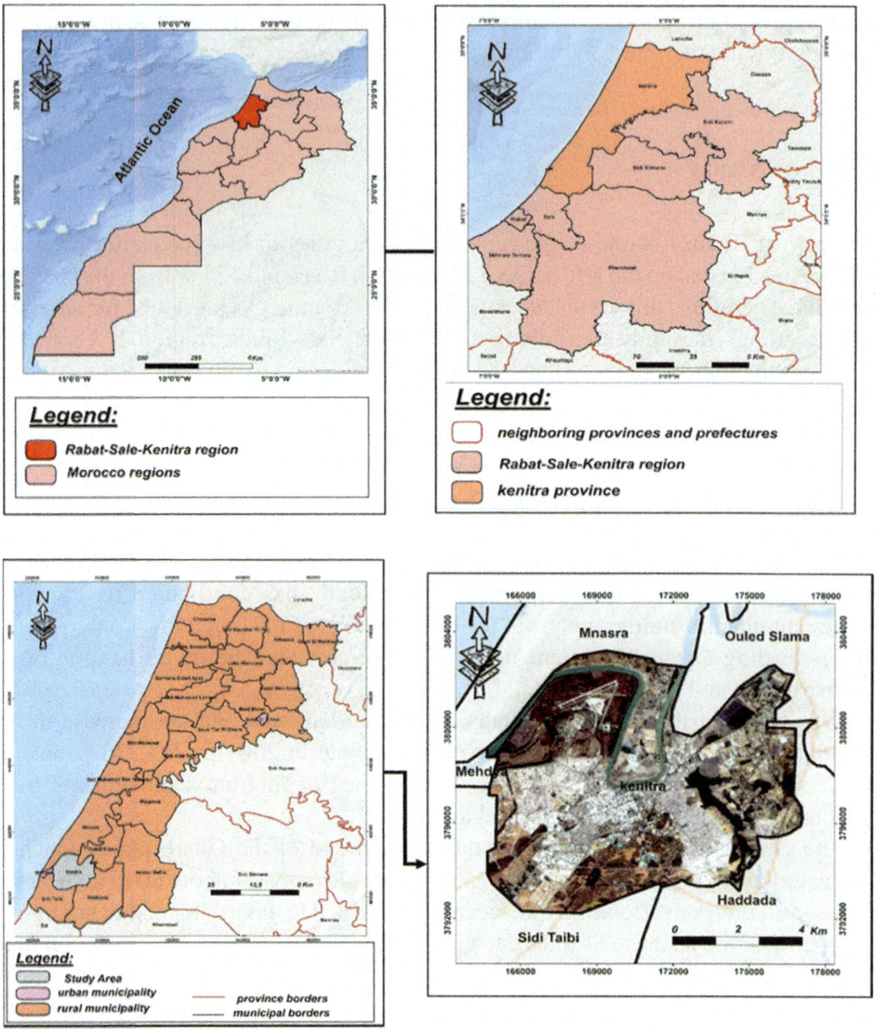

Fig. 11.1 Localizing the studied area at the national, regional, and local levels

Table 11.1 Characteristics of satellite imagery

Acquisition date	Spatial resolution of TIR cand (m)	Sensor	Satellite
08/08/1995	120	TM	LANDSAT 5
08/08/2021	100	OLI TIRS	LANDSAT 8

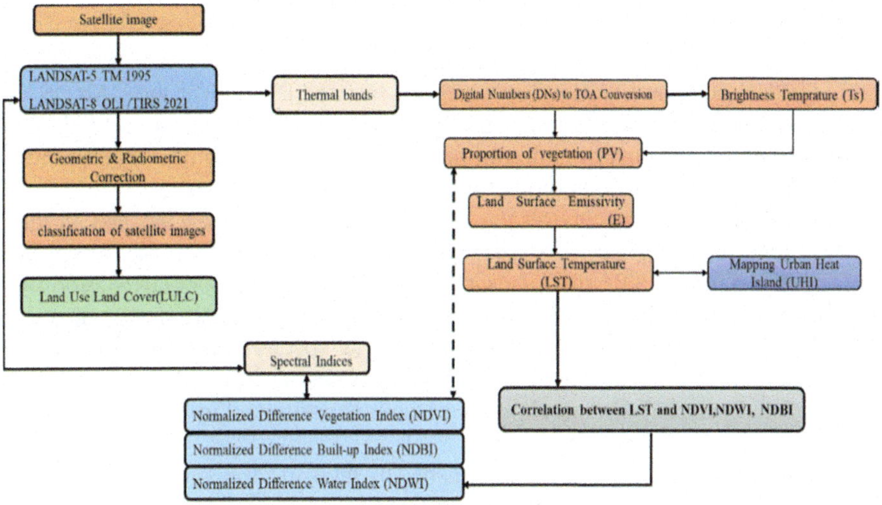

Fig. 11.2 The workflow of this study

and land used in the city. The sixth thermal band for the LANDSAT5TM sensor and the tenth band for the TIRS sensor of the LANDSAT8OLI/TIRS sensor were used for this purpose. The study utilized satellite images with cloud coverage below 10% (Avdan and Jovanovska 2016; Aboelnour and Engel 2018).

The remote sensing technique was used to analyze the satellite images using the ENVI 8.4 program and geographic information systems using the ArcGIS 10.8 program. In addition, a set of mathematical operations were performed to detect changes. The most important methods and tools used in the study were: Computation of Land Use Cover Change (LUCC), Land Surface Temperature (LST), Satellite Imagery and Urban heat island (UHI) Indices.

Computation of Land Use Cover Change (LUCC)

This study relied on the supervised classification method using a supervised algorithm called Maximum Likelihood Classification (MLC) to detect changes in land cover patterns, use, and areas (Li et al. 2022).Moreover, the four classes were extracted during the study years 1995 and 2021 as follows:

- Built-Up
- Bare Land
- vegetation Cover
- Water.

Land Surface Temperature (LST)

LST was derived from Landsat data using the infrared Band-6 for the LANDSAT5 TM sensor and the tenth band for the LANDSAT8 OLI/TIRS sensor and radiometrically corrected by converting digital number values to a thermal range (Band10/Band6) into a higher calculated spectral radiance (Liu et al. 2016).

The 10th band was chosen over the 11th band mainly because it has less air interference, less noise, and better sensor sensitivity when it comes to measuring land surface temperature (LST) with Landsat 8 Thermal Infrared Sensor (TIRS). Additionally, better atmospheric correction and more accurate differentiation of temperature differences are made possible by Band 10's shorter spectral range, leading to more accurate LST retrievals (Wang et al. 2015). The Thermal Infrared Sensor (TM) on Landsat 5 has selected the sixth band because it is positioned in a spectral window with minimal atmospheric interference.

Bands 6 and 10 are primarily distinguished by their spectral widths. Compared to Band 10, which has a spectral range of 0.59 μm, Band 6 has a wider range of 2 μm. Due to its broader spectral range, Band 6 is more susceptible to air interferences, especially those caused by vaporized water. As a result, Band 10 is typically seen to be more precise for LST computations (Kaiser et al. 2022).

The top of atmospheric (TOA) should be converted from spectral radiance to reflectance and brightness in the upper part of the atmosphere and then the land surface temperature (LST) is calculated using the following Eq. (11.1).

$$\mathrm{LST} = \mathrm{T}/((1 + (\lambda \times \mathrm{T}/\rho) ln \epsilon)) \tag{11.1}$$

LSTLand Surface Temperature.

T Brightness temperature (Band 10, 11, or 6 depending on the case)
λ Wavelength of the thermal band
σ Stefan-Boltzmann constant
P Constant value obtained from the formula P = h*c/s
e Spectral emissivity of land cover types.

Satellite Imagery Indices Used in the Study Normalized

Difference Vegetation Index (NDVI)

The vegetation cover or vegetation index is one of the most widely used indices worldwide. It is used to verify the status of vegetation cover by calculating the ratio of near-infrared (NIR) and red bands through the application (Mallick et al. 2008) of the mathematical Eq. (11.2) as well to detect its density, distribution, differentiation between vegetation, soil, and water. NDVI values range between − 1 and 1 +.

A positive value of NDVI indicates an increase in vegetation cover. Furthermore, the green vegetation index typically ranges from 0.2 to 0.8 and this index depends on the near infrared and red bands in calculating values related to vegetation cover, which highlights the color composition that combines red and near infrared. This is because vegetation has a high reflectivity in this band (Sall et al. 2013).

$$\mathrm{NDVI} = (\mathrm{NIR} - \mathrm{RED})/(\mathrm{NIR} + \mathrm{RED}) \tag{11.2}$$

where

NIR the reflectance value of near-infrared radiation
Red the reflectance value of visible red radiation.

The Normalized Difference Built-Up Index (NDBI)

NDBI can be calculated using Eq. (11.2). NDBI values range from − 1 to + 1, where the negative values correspond to the water bodies, the positive values correspond to urban areas, and the vegetation (Zheng et al. 2021) has a low value as determined through the application of Eq. (11.3).

$$\mathrm{NDBI} = (\mathrm{SWIR} - \mathrm{NIR})/(\mathrm{SWIR} + \mathrm{NIR}) \tag{11.3}$$

where

SWIR the reflectance value of short-wave infrared radiation
NIR the reflectance value of near-infrared radiation.

The Normalized Difference Water Index (NDWI):

On the one hand, the NDWI index is used to identify water bodies and monitor changes in water content. Water bodies absorb visible light and reflect a high amount of infrared radiation in the green wavelength range (60.0–52.0 μm) and very little in the near-infrared range (0–9076.0 μm). On the other hand, vegetation and soil have high reflectance values in the near-infrared wavelength range resulting in positive values for NDWI. Additionally, the areas with water bodies appear bright and have positive NDWI values, while green and built-up areas appear dark and have negative or zero values as calculated using the Eq. (11.4) (Xu 2006; Tan et al. 2020).

$$\mathrm{NDWI} = (\mathrm{GREEN} - \mathrm{NIR})/(\mathrm{GREEN} + \mathrm{NIR}) \tag{11.4}$$

where

GREEN the value of visible green light radiation
NIR the value of near-infrared radiation reflectance.

Urban Heat Island (UHI)

In this study, the UHI effect was extracted for the day by reference to Daily-average temperature. The urban heat island occurs when the temperature in the urban centre is above its environment (Moisa and Gemeda 2022). To compare the UHI change of different years, the UHI is calculated using the equation.

$$UHI = \frac{\text{LST} - \text{LSTmean}}{STD} \quad (11.5)$$

With

UHI urban heat island
LST_{mean} mean land surface temperature
STD standard deviation.

Results and Discussion

Identification of Ground Cover Types and Land Use

The current study revealed changes in the area during the period between 1995 and 2021. It was found that the main changes occurred in green areas, built or residential lands, and water areas, as shown in Table 11.2. The results of land classification for the city of Kenitra in 1995 showed that green areas represented the largest extension with an estimated area of about 74 km^2, accounting for 36%. However, the other types of land uses shared the rest of the area (barren areas with approximately 32.20 km^2, urban areas with approximately 24.25 km^2, and water areas with approximately 8.13 km^2). During this period, the city of Kenitra was developing gradually, as shown in Fig. 11.3.

As shown by the result classifications for 2021, the features of urban development and expansion increased by 15% at the expense of barren areas, which decreased by 5% with an estimated area of 4.87 km^2. The vegetation, consisting of green areas

Table 11.2 Area change in different LULC classes

LULC class	1995		2021	
	Area (in km^2)	Percentage (%)	Area (in km^2)	Percentage (%)
Built-Up	2,424,747,957	24	38,88,033,267	39
Bare land	3,220,071,888	32	27,32,368,347	27
Vegetation cover	3,574,364,126	36	27,07,134,612	27
Water	8,131,811,797	8	7,197,553,952	7

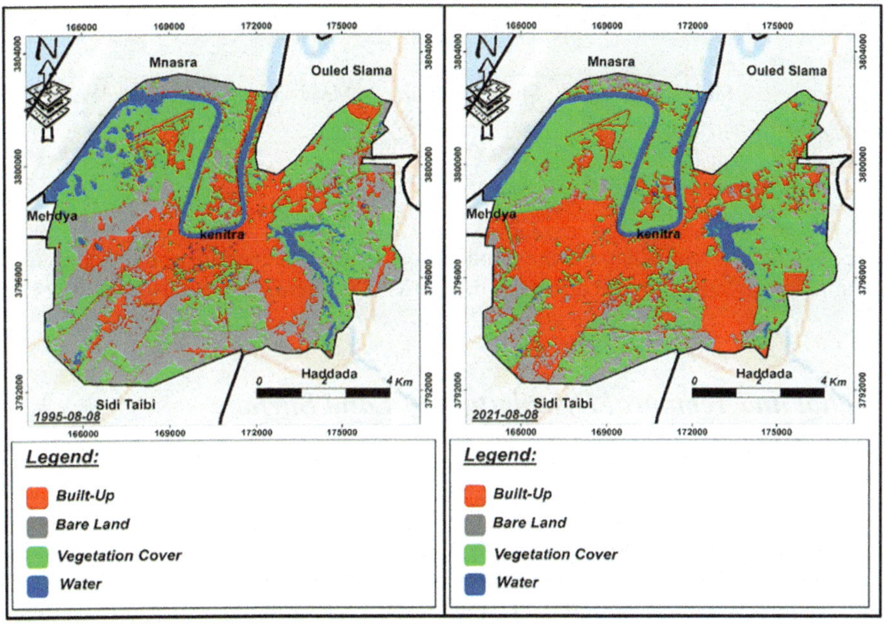

Fig. 11.3 LULC maps of the study area for the year 1995, 2021

and agricultural land, has decreased by 9%. It's estimated to be 8,67 km^2. This clearly indicates a lack of clear policies and strategies to guide and determine urban extensions in the city of Kenitra, to preserve and protect the declining vegetation cover and cultivate agricultural land.

The water category also saw a decline of 1% as shown in Fig. 11.3 which highlights the industrial activity that the city has witnessed in recent years. Also, this is with the inclusion of major industrial units in the free zone of Ouled Birrahmoun. All these factors have produced thousands of jobs, particularly in the automotive sector, making the city an important industrial center on both national and international levels. As a result, there has been a mass migration of job seekers, and sometimes entire families, to Kenitra. A large part of the workforce has settled throughout the city (in the residential neighborhood, the progress neighborhood, the OuledAref neighborhood, the AinSebaa neighborhood,…).

The city has also seen an improvement in its road transport networks with the completion of the highway and the upgrading of the road that connects Kenitra to Rabat as well as the unification of transport by rail. The high-speed TGV train link has also facilitated the movement of residents from Kenitra to their workplaces in either Rabat or Casablanca.

The study evaluated the accuracy of Land Use/Land Cover (LULC) categories, yielding overall accuracy assessment results of 81.6% for 1995 and 97% for 2021. Additionally, the Kappa statistics assessment produced values of 0.81 for 1995 and

Table 11.3 Statistical results of categorized LST C of kenitra in 1995, 2021

Date	LST			
	Maximum value	Minimum value	Mean value	Standard deviation
08/08/1995	36.8	16.1	30.9	2.87
08/08/2021	40.5	22	31.9	2.71

0.93 for 2021. These Kappa values, indicating agreement beyond chance, suggest a notable improvement in the accuracy of LULC classification from 1995 to 2021.

Spatial and Temporal Distribution of Land Surface Temperatures

According to the results obtained based on two satellite images captured in different years for the same location (2021–1995), it is evident from the data represented in Table 11.3 and Fig. 11.4 that in summer: the highest land surface temperature (LST) value recorded in 2021 was 40.5 °C, the lowest LST value was 22 °C. The highest and lowest LST value recorded in 1995, 36.8 °C, and 16.1 °C respectively. The maximum land surface temperature in the city of Kenitra increased by 3.7 °C during the period 1995–2021, while the average LST increased by 1 °C. These differences are attributed to the weather conditions observed during this period 1995 and 2021 in the summer.

High LST values are found in all densely populated urban areas, tourist areas, and industrial areas, especially in barren areas, as shown in the LST maps. The continuous increase in land surface temperature in the city of Kenitra during the study period is consistent with the above findings, which confirm that the area studied experienced significant urban growth and a significant reduction in vegetation, all of which resulted in a significant increase in land surface temperature in the area studied.

This indicates a high absorption coefficient for urban surfaces represented by high-density residential neighborhoods, commercial centers, connected cement spaces, and industrial areas.

Analysis of Land Cover Indicators for the City of Kenitra During the Study Periods

Normalized Difference Vegetation Index (NDVI)

The NDVI values were extracted as shown in Table 11.4 and Fig. 11.5, where the maximum and minimum values during the year 1995 ranged between 0.395 and

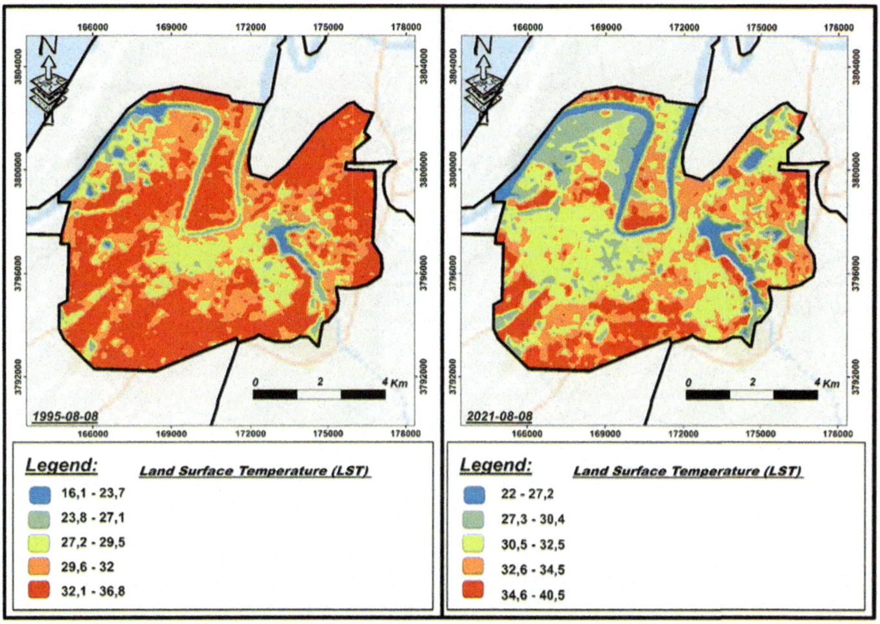

Fig. 11.4 Land surface temperature (LST) maps of kenitra in 1995, 2021

Table 11.4 Statistical results of categorized NDVI of kenitra in 1995, 2021

Date	NDVI			
	Maximum value	Minimum value	Mean value	Standard deviation
08/08/1995	0.679	– 0.395	0.06	0.102
08/08/2021	0.62	– 0.26	0.16	0.1

0.679, while during the year 2021, the values ranged between – 0.26 and 0.62. The maximum NDVI values in the city of Kenitra decreased by 0.05 during the study years 1995 and 2021, and the average NDVI decreased by 0.1. This indicates a significant difference in values and a change in the form of vegetation cover, with a decrease in green color, density and spread, as mentioned in Table 11.2, where the vegetation cover decreased by 9% in 2021.

The Normalized Difference Built-Up Index (NDBI)

Table 11.5 and Fig. 11.6 illustrate the distribution of NDBI values in the city of Kenitra during the study period. The maximum and minimum values ranged between 0.413 and 0.217 in 1995, and between 0.503 and 0.413 in 2021. The maximum value of the NDBI indicator increased by 0.29 during the study years 1995 and 2021, and the

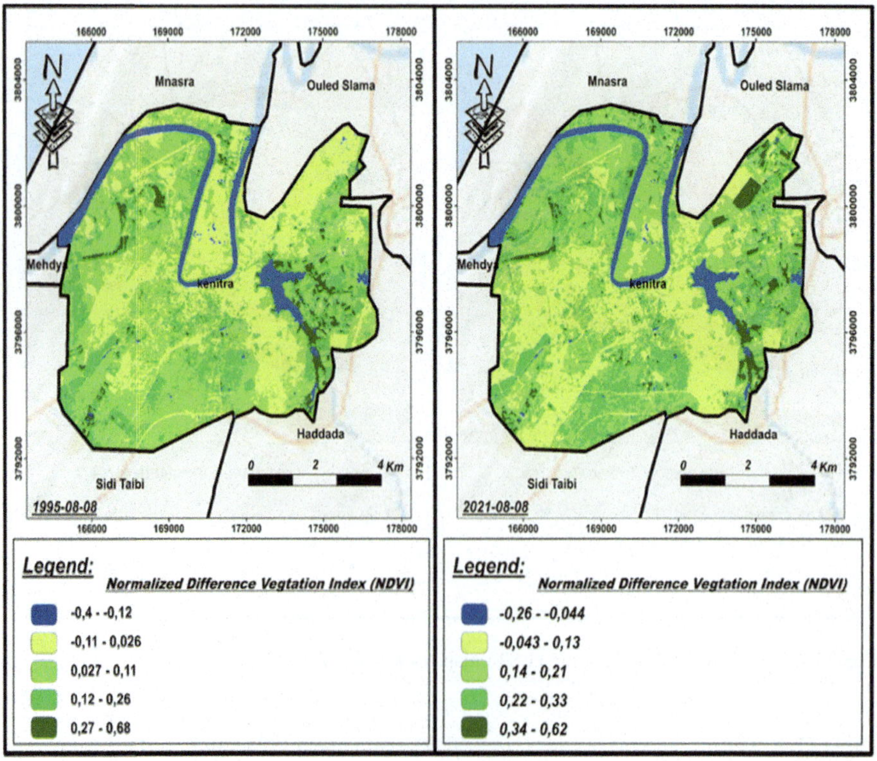

Fig. 11.5 Spatial distribution of NDVI of 1995, 2021

average NDBI increased by 0.286. This indicates the urban changes that occurred in the city during the study period, where the coverage of built-up areas increased at the expense of bare lands and vegetation cover.

However, this indicator is not as good as NDVI, given the overlap between NDVI and NDBI, which has more high-than-low implications for buildings, and the fact that the indicator cannot separate buildings in a pure manner because of the overlap of spectral reflections between buildings and arid areas that are rather poor with vegetation cover (Muhaimin et al. 2022).

Table 11.5 Statistical results of categorized NDBI of kenitra in 1995, 2021

Date	NDBI			
	Maximum value	Minimum value	Mean value	Standard deviation
08/08/1995	0.217	– 0.413	0.011	0.082
08/08/2021	0.503	– 0.4	0.284	0.105

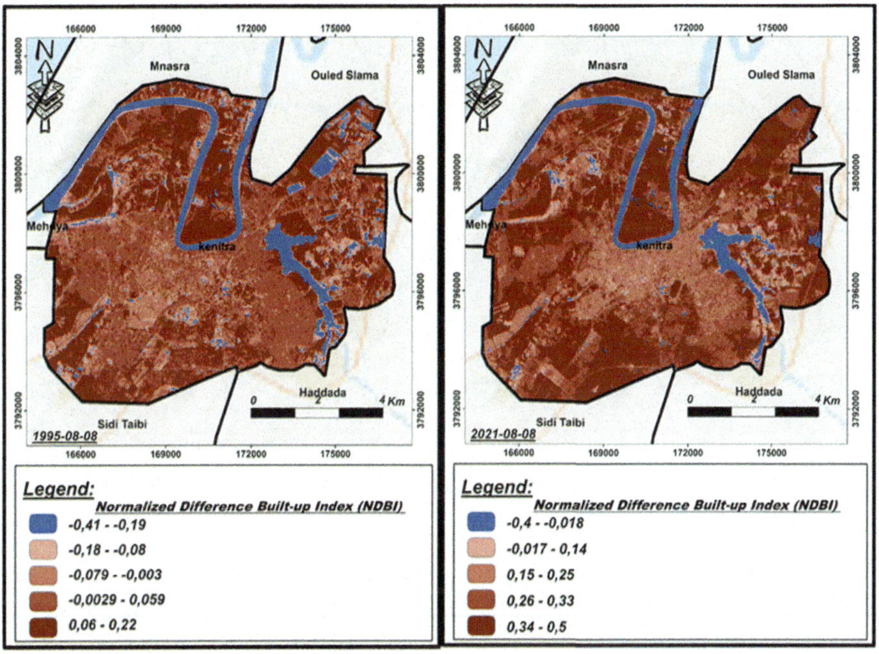

Fig. 11.6 Spatial distribution of NDBI of 1995, 2021

Table 11.6 Statistical results of categorized NDWI of kenitra in 1995, 2021

Date	NDWI			
	Maximum value	Minimum value	Mean value	Standard deviation
08/08/1995	0.458	− 0.64	− 0.182	0.112
08/08/2021	0.303	− 0.557	− 0.197	0.105

Normalized Different Water Index (NDWI)

Table 11.6 and Fig. 11.7 show the NDWI values, with maximum and minimum values ranging from 0.64 to 0.458 in 1995 and 0.557 to 0.303 in 2021. The maximum NDWI values in Kenitra decreased by 0.155 between 1995 and 2021, and the average NDWI decreased by − 0.015, due to the impact of climate on the region.

Relationship Between Spectral Indices and Surface Temperature

To establish a connection between surface temperature and land cover indices in Kenitra, a regression analysis and linear correlation were conducted using three land cover indices: NDVI, NDWI, and LST. The findings of the statistical analysis and

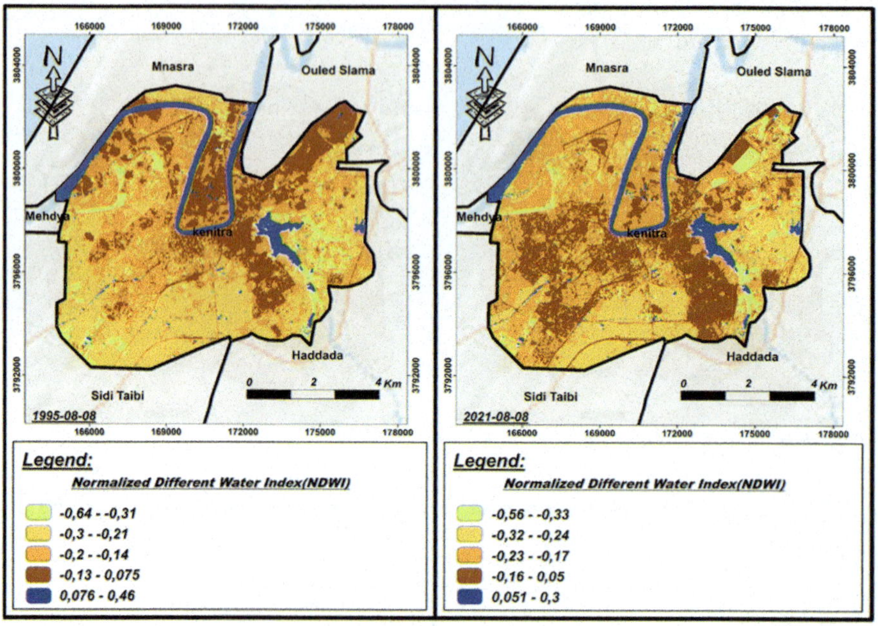

Fig. 11.7 Spatial distribution of NDWI of 1995, 2021

the relationship between surface temperature and land cover indices during the study years 1995 and 2021 are displayed in Fig. 11.8.

The outcomes of the study indicate a negative correlation between LST and both NDVI and NDWI. This implies that the reduction in water surfaces and vegetation cover in the area studied has led to an increase in surface temperature, as depicted in the table by the correlation coefficients of $R^2 = 0.1263$ and $R^2 = 0.0692$. However, the correlation coefficient between LST and NDWI was insignificant, at $R^2 = 2 \times 10^{-6}$.

Overall, the study's results suggest that the decrease in vegetation cover and water surfaces in the study area is the primary factor behind the increase in surface temperature.

Figure 11.8 also shows a positive correlation between LST and NDBI with correlation coefficients of $R^2 = 0.509$ and $R^2 = 0.1713$ during the study years 1995 and 2021, indicating that built-up areas generate many changes in surface temperature and are the main contributor to the appearance of urban heat islands.

The Urban Heat Island of the City of Kenitra

The results of the study showed that the spatial pattern of UHI increased during the study years 1995 and 2021 from the city center in all directions. This spatial and temporal difference was associated with urban expansion, where the maximum value

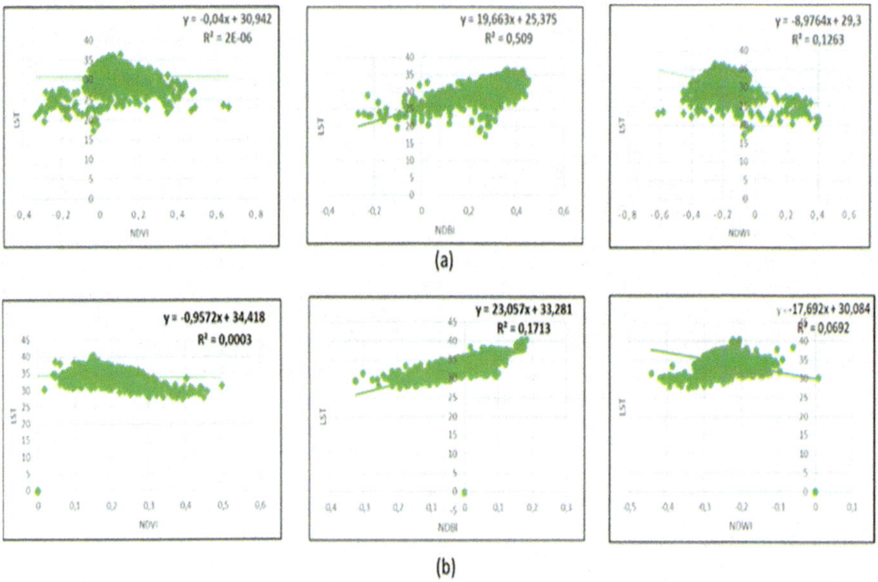

Fig. 11.8 Regression correlation between LST, NDVI, NDWI and NDBI. **a** 1995, **b** 2021

of urban heat island (UHI) increased from 23.60 °C in 1995 to 26.70 °C in 2021, with an increase rate of 3.1 °C. The minimum value also increased from 16.37 °C in 1995 to 19.88 °C in 2021, with an increase rate of 3.51 °C (Fig. 11.9).

The most significant changes have occurred in barren lands, residential areas, and green spaces, highlighting the conversion of bare lands to residential and industrial uses, resulting in massive structures on non-reflective surfaces. The use of materials like asphalt and concrete, which are non-heat reflective, and human activities such as greenhouse gas emissions have contributed to the increase in UHI in the urban environment, as illustrated in the picture (Fig. 11.10).

Furthermore, the study found that the surface temperature in the industrial area of the city exceeds 40 °C, suggesting that iron and steel factories have played a significant role in increasing the earth's surface temperature.

Kenitra has a long history of manufacturing, and the post-independence period witnessed the establishment of numerous industrial facilities, including the old industrial neighborhood, municipal industrial neighborhood, alfouarat-issam industrial neighborhood, Sakenia neighborhood, and the Bir Rami industrial area. The UHI effect has implications for the urban thermal comfort of the city's inhabitants and the overall climate of the city.

In industrial cities, pollutant levels of greenhouse gases rise, leading to an increase in temperatures, a phenomenon known as thermal anomalies. Consequently, the heat islands expand into industrial areas and may extend to encompass large surrounding areas. As industrial processes and the construction of industrial cities intensify, the size of these warm islands in the city increases, and vice versa. Urban planning often

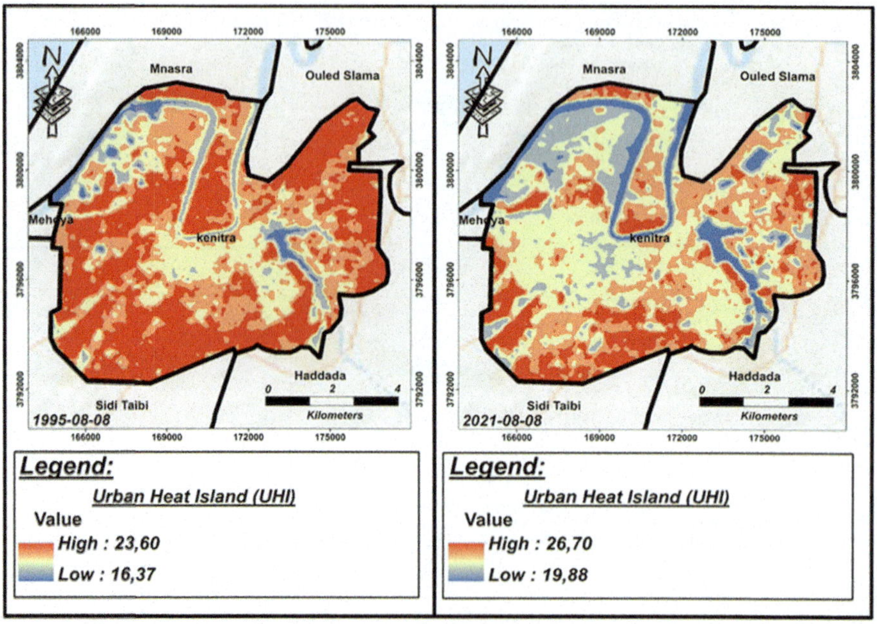

Fig. 11.9 UHI effect map of kenitra city between 1990 and 2022

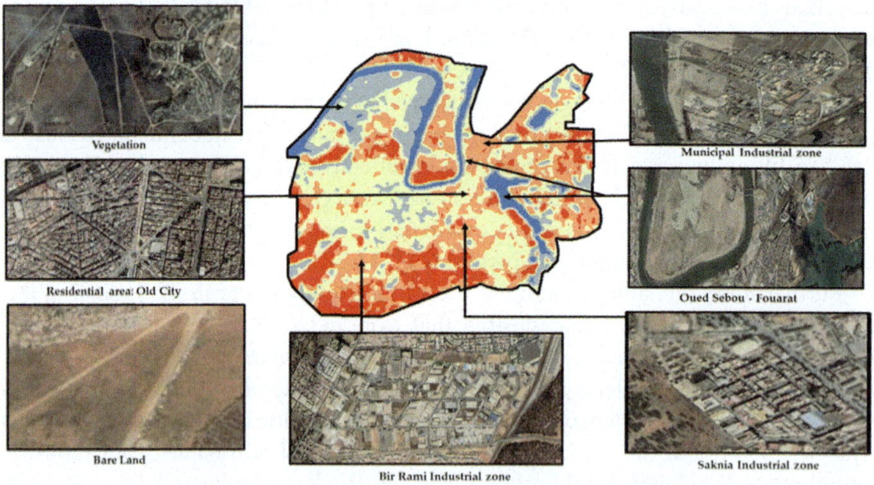

Fig. 11.10 Detailed views of the different land use classes of the conurbation

reveals the poor placement of industrial zones, commonly situated amidst residential areas.

Conclusions

This research utilizes remote sensing and geographic information systems to analyze the distribution of land surface temperature in urban areas. It has been observed that land use types, such as residential and industrial development, as well as vacant land, contribute to an increase in land surface temperature, while vegetation and water cover lead to a decrease. The study recommends urban planning strategies, such as creating green spaces in urban areas and limiting urban expansion at the expense of vegetation cover, to mitigate the impact of urban heat islands on human thermal comfort. Integrating remote sensing data into geographic information systems can be a powerful tool for planning and managing spatial data analysis to make informed decisions. The study provides valuable information about the relationship between land cover types and land surface temperature. This information can assist in identifying suitable locations for industrial projects, evaluating potential environmental impacts, and developing mitigation measures. By understanding the environment and utilizing appropriate technologies, industrial planners can contribute to a more sustainable future.

References

Aboelnour M, Engel BA (2018) Application of remote sensing techniques and geographic information systems to analyze land surface temperature in response to land use/land cover change in Greater Cairo Region, Egypt. J Geogr Inform Syst 10(1). https://doi.org/10.4236/jgis.2018.101003

Avdan U, Jovanovska G (2016) Algorithm for automated mapping of land surface temperature using LANDSAT 8 satellite data. J Sens 2016:e1480307. https://doi.org/10.1155/2016/1480307

Azmi R, Tekouabou koumetio CS, Diop EB, Chenal J (2021) Exploring the relationship between urban form and land surface temperature (LST) in a semi-arid region case study of Ben Guerir city—Morocco. Environ Challenges 5:100229. https://doi.org/10.1016/j.envc.2021.100229

Bahi H, Rhinane H, Bensalmia A, Fehrenbach U, Scherer D (2016) Effects of urbanization and seasonal cycle on the surface urban heat island patterns in the coastal growing cities: a case study of Casablanca, Morocco. Remote Sens 8(10). https://doi.org/10.3390/rs8100829

Dibs H, Alnajjar H (2013) Maximum likelihood for land-use/land-cover mapping and change detection using landsat satellite images: a case study “South Of Johor”. Int J Comput Eng Res 3:26–33

Gadal S (2008) Télédétections thermiques infrarouges des concentrations urbaines au Maroc. Cybergeo: Europ J Geogr. https://doi.org/10.4000/cybergeo.18592.

Garouani ME, Amyay M, Lahrach A, Oulidi HJ (2021) Land surface temperature in response to land use/cover change based on remote sensing data and GIS techniques: application to Saïss Plain, Morocco. J Ecol Eng 22(7):100–112. https://doi.org/10.12911/22998993/139065

Huang S, Tang L, Hupy JP, Wang Y, Shao G (2021) A commentary review on the use of normalized difference vegetation index (NDVI) in the era of popular remote sensing. J Res 32(1):1–6. https://doi.org/10.1007/s11676-020-01155-1

Kafy A-A et al (2020) Impact of LULC changes on LST in Rajshahi District of Bangladesh: a remote sensing approach. J Geograph Stud 3:11–23. https://doi.org/10.21523/gcj5.19030102

Kaiser EA et al (2022) Spatiotemporal Influences of LULC changes on land surface temperature in rapid urbanization area by using Landsat-TM and TIRS images. Atmosphere 13(3). https://doi.org/10.3390/atmos13030460.

Li X, Stringer LC, Dallimer M (2022) The Impacts of urbanisation and climate change on the urban thermal environment in Africa. Climate 10(11). https://doi.org/10.3390/cli10110164

Liu Z, Yao Z, Wang R (2016) Assessing methods of identifying open water bodies using Landsat 8 OLI imagery. Environ Earth Sci 75(10):873. https://doi.org/10.1007/s12665-016-5686-2

Mallick J, Kant Y, Bd B (2008) Estimation of land surface temperature over Delhi using Landsat– & ETM+. J Ind Geophys Union 12:131–140

Moisa MB, Gemeda DO (2022) Assessment of urban thermal field variance index and thermal comfort level of Addis Ababa metropolitan city, Ethiopia. Heliyon 8(8):e10185. https://doi.org/10.1016/j.heliyon.2022.e10185

Muhaimin M, Fitriani D, Adyatma S, Arisanty D (2022) Mapping build-up area density using normalized difference built-up index (NDBI) and urban index (UI) wetland in the city Banjarmasin. IOP Conf Ser Earth Environ Sci 1089(1):012036. https://doi.org/10.1088/1755-1315/1089/1/012036

Sall B, Jenkins MW, Pushnik J (2013) Retrospective analysis of two northern California wild-land fires via Landsat five satellite imagery and normalized difference vegetation index (NDVI). Open J Ecol 3(4). https://doi.org/10.4236/oje.2013.34036

Tan J, Yu D, Li Q, Tan X, Zhou W (2020) Spatial relationship between land-use/land-cover change and land surface temperature in the Dongting Lake area, China. Sci Rep 10(1). https://doi.org/10.1038/s41598-020-66168-6

Wang S, He L, Hu W (2015) A temperature and emissivity separation algorithm for Landsat-8 thermal infrared sensor data. Remote Sens 7(8). https://doi.org/10.3390/rs70809904

Xu H (2006) Modification of normalized difference water index (NDWI) to enhance open water features in remotely sensed imagery. Int J Remote Sens 27:3025–3033. https://doi.org/10.1080/01431160600589179

Zheng Y, Tang L, Wang H (2021) An improved approach for monitoring urban built-up areas by combining NPP-VIIRS nighttime light, NDVI, NDWI, and NDBI. J Clean Prod 328:129488. https://doi.org/10.1016/j.jclepro.2021.129488

Chapter 12
The Effect of Incorporating Alfa Fibers on Physical, Mechanical, and Durability Properties of Compressed Stabilized Earth Blocks

Reda Sadouri, Youssef Cherradi, and Mustafa Benyoucef

Abstract This paper investigates the effect of integrating Alfa fibers into compressed earth blocks (CEBs) stabilized with varying Portland cement contents. CEB composites were manufactured with earth stabilized using different cement contents (5 and 10% by weight) and Alfa fibers reinforcement (0–0.4% by weight), compressed at 10 MPa with a compaction loading press. After 28 days of drying, the CEBs underwent diverse experimental tests to evaluate their physical, mechanical, and durability properties. The findings indicated that incorporating fibers led to a diminution in unit weight, ultrasonic pulse velocity, and dry compressive strength. Despite the drop in mechanical strength, CEBs with lower cement (5%) and higher fiber content (0.4%) demonstrated satisfactory erosion resistance, which could play a crucial role in areas prone to extreme weather events (floods and storms). According to this study, this material has the potential as a promising composite for building materials and reducing the need for cement, which is deemed a major contributor to greenhouse gas emissions in the construction industry. The results have implications for affordable housing solutions and offer insightful information about sustainable building materials.

Keywords Compressed earth blocks · Alfa fiber · Dynamic modulus · Compressive strength · Erosion resistance

---

R. Sadouri (✉) · Y. Cherradi · M. Benyoucef
LRDDS, Department of Applied Physics, Faculty of Sciences and Technics, Cadi Ayyad University, 40000 Marrakesh, Morocco
e-mail: r.sadouri@uca.ma

Y. Cherradi
e-mail: y.cherradi@emsi.ma

M. Benyoucef
e-mail: m.benyoucef@uca.ac.ma

Y. Cherradi
LAMIGEP, Moroccan School of Engineering Sciences, Marrakech, Maroc

K. Kahime et al. (eds.), *Climate Change Effects and Sustainability Needs*, Springer Climate, https://doi.org/10.1007/978-3-031-59603-2_12

Introduction

To address climate change, promote affordable and clean energy, and create sustainable cities and communities, countries have pledged to implement measures about the building sector under the Sustainable Development Goals (SDGs). The building industry has become the main obstacle on the road to achieving these global goals and protecting our planet (Ness and Xing 2017). Approximately 33% of CO_2 emissions, 40% of the world's energy consumption, and 40% of all trash output are attributable to the building industry (Ness and Xing 2017). Such situations have motivated researchers around the globe to study the potential of stabilized soil using affordable alternatives with a low environmental impact (Danso et al. 2015).

Meanwhile, it is widely acknowledged that there is a shortage of reasonably priced housing on the planet. More than 60 million shelters are needed in Africa to face the housing scarcity. Efforts are being made to encourage the construction of new houses in low-income areas by using available and local construction materials. Earthen buildings structures have been used for ancient time and even today, approximately 40% of the world's population live and work in earthen buildings (Mateus et al. 2019; Minke 2013). Earth is an eco-friendly, cheap, widely abundant material that can help to promote living in comfort and decrease environmental issues.

Climate change has emerged as a major global threat and a difficult environmental and economic issue in recent years (Miller et al. 2018; Miller and Moore 2020; Fawzy et al. 2020). There were 315 natural disasters worldwide in 2018; most of them were brought on by climate change. Storms, floods, wildfires, and droughts caused approximately 70 million people to be touched and caused 131.7 billion dollars in economic losses; storms, floods, wildfires, and droughts made up about 93% of the total (Fawzy et al. 2020). Countries will be compelled to turn back to clean, renewable resources in the form of renewable energies and give up using petroleum products as a source of energy, as well as to use renewable raw materials in fundamental industries including the building materials sector, in light of the threat to the global economy and environment. In this regard, the building industry needs to change the way it constructs buildings in order to increase the energy efficiency of both new and existing structures while also providing novel materials that satisfy the evolving needs of users with regard to comfort, health effects, and the environment. Recently, the scientific community has been able to shed light on soil as building materials and is working to make them promising, clean and economical alternatives to fired brick and cement materials.

Compressed earth blocks (CEBs) emerged as a popular earth construction technique in the 1950s. Building upon the traditional Adobe method, they involve the same production steps but with significantly higher compaction pressure. Additionally, various stabilization techniques (mechanical, physical, and chemical) have been implemented to enhance their strength and lifespan. However, earth blocks often face limitations in terms of durability, erosion resistance, and overall strength (Islam et al. 2020). Many regions worldwide are vulnerable to natural disasters such as monsoons, floods, storm surges, and tsunamis. These events often cause widespread damage and devastation, including the collapse of buildings. Such collapses not only result in loss of life and property but also significantly hinder progress towards the Sustainable Development Goals (SDGs) (Islam et al. 2020). However, there has been a rebirth of interest in ecologically friendly, healthful earthbased construction materials as a result of these materials' poor thermal performance, increased expense, and particularly their increased environmental effect (Alam et al. 2015; Arrigoni et al. 2017; Burroughs 2010; Eires et al. 2017). Several studies on CEBs have demonstrated that adding Portland cement and other additives to earth blocks increases its mechanical qualities and water resistance (Islam et al. 2020; Abhilash et al. 2020; Mostafa and Uddin 2016; Taallah et al. 2014; Zak et al. 2016; Nagaraj et al. 2014).

Many authors have concentrated on the properties of stabilized CEBs (Mostafa and Uddin 2016; Zak et al. 2016; Khedari et al. 2005; Kriker et al. 2005; Taallah and Guettala 2016; Saghrouni et al. 2020; Limam et al. 2016; Panesar and Shindman 2012; Adam and Jones 1995). The compressive strength of CEBs is influenced by the test method used (Mostafa and Uddin 2016) and increases with the increase in binder content (Khedari et al. 2005). The study of the impact of the curing period in CEBs stabilized with cement has shown that 28 days of curing provides the largest value of compressive strength (Kriker et al. 2005). Other research studies (Alam et al. 2015; Arrigoni et al. 2017; Burroughs 2010; Eires et al. 2017; Mostafa and Uddin 2016; Taallah et al. 2014; Zak et al. 2016) have been conducted on cement, a common and contemporary chemical stabilizer that is typically combined with fibers and other additives. However, because Portland cement requires energy to produce, using it increases the embodied energy of the stabilized CEBs. Furthermore, the ductility, toughness, and tensile strength were not enhanced by the use of chemical stabilizers (Zare et al. 2020; Segetin et al. 2007). In addition, because of the chemical reaction that turns earth construction material into fake stones, it creates a problem with recycling (Sujatha and Devi 2018). Numerous studies have examined how chemical stabilization affects water absorption and long-term strength (Nagaraj et al. 2014). It has been observed that as time passes, compressive strength rises while water absorption falls. According to Bahar et al. (2004), the ideal way to stabilize the blocks may be a combination of mechanical compaction and chemical stabilization up to a certain point. This would result in improved durability and compressive strength at an affordable price.

Additional research on thermal characteristics has shown that bricks stabilized with cement have a high degree of thermal conductivity (Adam and Jones 1995). Moreover, heat conductivity rises with increasing cement and lime content (Khedari et al. 2005; Zhang et al. 2017).

Naturally fiber-reinforced biopolymer materials will be the future of "green composites," as they might potentially address sustainability issues including pollution in the environment and resource scarcity. Traditionally, since ancient times, natural fibers have been incorporated into earthen building materials to improve composite performance, including ductility, toughness, tensile strength, and durability, as well as increasing its strength (Islam et al. 2020; Zare et al. 2020; Segetin et al. 2007; Sujatha and Devi 2018) while reducing shrinkage and cracks. Plant fibers have garnered special attention for use as building materials because of their ability to replace synthetic fibers and because of their worldwide abundance (Zare et al. 2020; Segetin et al. 2007). Additionally, some research has found that the thermal conductivity of CEBs decreases as the fiber content increases (Khedari et al. 2005; Kriker et al. 2005; Garrouri et al. 2022; Omrani et al. 2020; Williams et al. 2016; Sadouri et al. 2024). According to the literature (Achenza and Fenu 2006), The capacity of the matrix to fill the pore network and glue the grains together results in an improvement in mechanical strength with an increase in fiber content while maintaining a low unit weight. Lastly, fibers reduce the density of the earth composite, which has several advantages such as improved protection against earthquakes due to a decrease in heat conductivity and inertial forces (Islam and Iwashita 2010; Gandia et al. 2019; Binici et al. 2007).

The goal of this experimental study is to determine whether it is feasible to produce compressed earth blocks with improved properties by adding Alfa fibers as a reinforcement and Portland cement as a stabilizer. The CEBs were stabilized using Portland cement in weight fractions of 5, and 10%. The reinforcing fibers were added at different ratios, specifically 0, 0.10, 0.20, 0.30, and 0.4%. The resultant CEB composites were experimentally characterized in order to examine their density, ultrasonic pulse velocity, elastic dynamic modulus dry compressive strength, durability characteristics (erosion resistance). To the best of our knowledge, no literature has described this particular formulation for compressed earth blocks, which uses Alfa fibers as a raw material reinforcement and Portland cement as a stabilizer.

Materials and Experimental Testing

Materials

Earth

The natural soil used in this study as shown in Fig. 12.1 was collected from Harbil area located 20 km northwest of Marrakesh. The earth was extracted from a layer of 1 m under the ground to avoid organic content. This soil was selected because of its abundance and availability in the area. The granulometric size distribution (GSD) was determined following ASTM standard D422 (2007) (Binici et al. 2007): clay: < 0.002 mm; silt: 0.002–0.08 mm; sand: 0.08–2 mm; and gravel: > 2 mm. Atterberg's test (the plastic limit, index plasticity and liquid limit) was performed using ASTM D4318-10 (2007, 2010). The methylene blue value of this soil was determined according to the NM 13.1.178 standard. The optimum moisture content for the soil was obtained according to the standard Proctor test (ASTM Committee D-18 on Soil and Rock 2007).

The properties of the soil used in this study are shown in Table 12.1a. The soil was constituted of 14.1% clay, 35.4% silt, 30.7% sand and 19.8% gravel. The Atterberg limit test indicated that this soil had a liquid limit WL = 29%, plastic limit WP =

Fig. 12.1 The earth used in this work

21% and a plasticity index PI = 8%, which agreed within the recommended zone limit by the French Standards XP P 13-901 (Afnor 2001). It had an absolute density of 2400 kg/m^3, and bulk density 1300 kg/m^3 and a blue methylene value MBV (g/100 g) = 1.25. Therefore, the soil could be classified as type A1, in close agreement with the XP P 13-901 standard for the preparation of compressed earth blocks (Afnor 2001).

The chemical composition of the used raw soil shown in Table 12.1b was carried out X-ray fluorescence (XRF), and mineral composition was determined using X-ray diffraction (XRD). The results indicate that the soil consisted essentially of silica, aluminum and calcium carbonate. In addition, it also contained significant quantities of carbon (8%) and magnesium (4%) and a small amount of fluorine (0.87%).

Table 12.1 (a) Geotechnical characteristics of soil. (b) Chemical and mineralogical characteristics of soil

(a)

Grain size distribution	Atterberg limits	Methylene blue value	Absolute density
Clay < 2 μm: 14.1% Silt (2–63 μm): 35.40% Sand (0.063–2 mm): 30.70% Gravel (> 2): 19.80%	Liquid limit, WL = 29% Plastic limit, WP = 21% Plasticity index, PI = 8%	MBV (g/100 g) = 1.25	2400 kg/m^3

(b)

Oxides	C K	O K	F K	Mg K	Al K	Si K	Total
Wt%	8.39	49.28	0.87	4.39	11.19	25.85	100.01

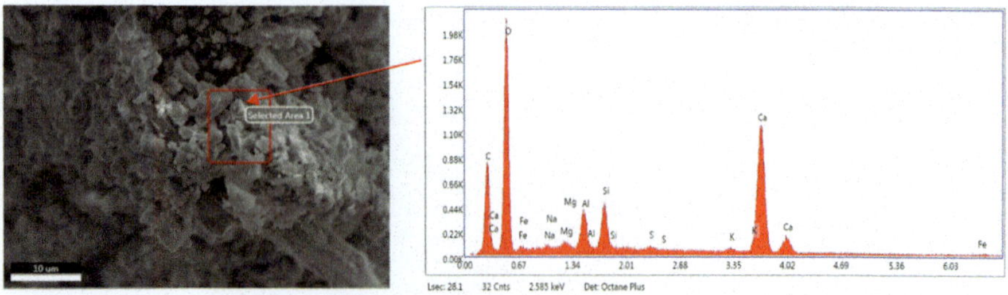

Fig. 12.2 SEM micrographs/energy-dispersive X-ray (EDX) of cement

Cement

Portland cement was used in this study (CEM CPJ 45), created by mixing clinker and gypsum with one or more secondary ingredient, such as fillers, pouzzolane, or fly ash, as specified in Moroccan Norm NM 10.1.004. The representative energy dispersive X-ray (EDX) spectra of Portland cement are illustrated in Fig. 12.2. The results indicate the peaks of this binder. It can be noticed from the spectra of EDX that the cement used in this study was mostly composed of oxygen (O), calcium (Ca), carbon (C) and silicon (Si) mixed with traces of aluminum (Al), magnesium (Mg), sodium (Na), sulfur (S), iron (Fe) and potassium (K).

Alfa Fibers

The natural fibers used in this work were from the Alfa plant also known as esparto grass. It belongs to Stippa tenacissima family. Its stems have a length of approximately 2 m, with diameter of 2–3 mm (Ajouguim et al. 2019), thrived in the dry and arid regions of North Africa, particularly Morocco, Algeria, and Tunisia and the South of Europe. The Alfa plant grew in abundance in many regions, essentially in the north-eastern of Morocco. The fibers used in this study were harvested from the region of "Oujda". Before their use in the composite, the fibers were cleaned to remove any dust and impurities deposited on the surface, then they were cut at lengths between (1 and 3 cm). Figure 12.3 shows the cut Alfa fiber studied in this study.

According to Garrouri et al. (2022), Alfa fibers are mainly composed of cellulose (39.50%), hemicellulose (27.60%), lignin (19.50%), and extractible element (13.31%) as shown in Table 12.2a. Table 12.2b lists the properties of the fibers determined experimentally according to ASTM C1557 (2003). Figure 12.2 shows the structure of the Alfa plant observed through scanning electronic microscopy (SEM). Based on longitudinal stem observation, the SEM observations show that Alfa stems have a hairless and soft appearance with numerous dusts and impurities (wax, fat, etc.) that are located on the surface (Fig. 12.4b). Additionally, the SEM of the fiber cross section indicates that this plant has a cellular structure with pores (Fig. 12.4a).

Fig. 12.3 Cut and cleaned Alfa fibers used in this work

This type of vegetal fiber's void network morphology provides promise for creating composites with improved thermal performance and low unit weight.

Table 12.2 (a) Chemical composition of Alfa fibers. (b) Properties of Alfa fibers

(a)			
Cellulose (%)	Hemicellulose (%)	Lignin (%)	Extractible element (%)
39.50	27.60	19.50	13.31
(b)			
Density (g/cm^3)	Young's modulus (GPa)	Stress at break (MPa)	Strain at break (%)
1.40	21.5	247	1.96

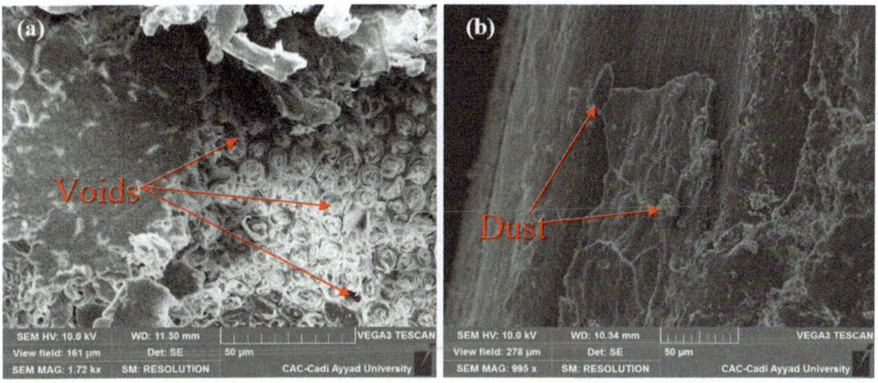

Fig. 12.4 **a** SEM micrograph of a cross-section of the stem of a Juncus plant. **b** SEM of a longitudinal section of the stem of a Alfa plant

Table 12.3 The proportion of mixtures of CEB

	Composition for (5% cement + Different fiber contents)	Composition for (10% cement + Different fiber contents)
Soil in relation to the dry soil (%)	100	100
Cement in relation to the dry soil (%)	5	10
Fibers in relation to the dry soil (%)	0	0
	0.10	0.10
	0.20	0.20
	0.30	0.30
	0.40	0.40
Water in relation to the dry soil	13.50	15.50

Sample Preparation

Different compositions of CEBs samples were prepared with different contents of fibers, as shown in Table 12.3. The preparation of specimens consisted of first mixing dry soil and cement using a concrete mixer for 5 min, and then gradually adding the dry fibers to the mix until a homogeneous color of mixture was obtained, then water was added to the mixture that was continuously being mixed for 5 min, thus accounting for the water demand for cement to estimate its effect on the stabilization of the CEBs. Afterward, the mixture was placed into steel molds dimension (295 mm × 145 mm × 95 mm) and then, compacted with static loading pressure of 10 MPa. After demolding, the drying of the samples took place in normal laboratory conditions at 20 ± 2 °C with a relative humidity of 98% for approximately 28 days until constant weight, the steps for the production of the CEBs are presented in Fig. 12.5. And the curing conditions to avoid unwanted waiting are shown in Fig. 12.6, in which the blocks were cured at 105 °C for 24 h before testing.

Experimental Testing

The physical properties of the CEB samples containing different contents of Alfa fibers, including the dry unit weight, as concluded by means of dimensional measurements and weighing after 28 days of curing, were established.

Fig. 12.5 Main steps in the production process of CEBs, including **1** after the mixture of the composite, **2** the filling of the mold, and **3** the block after compacting stress

Fig. 12.6 Cured CEB samples in the oven

Ultrasonic Testing

The ultrasonic pulse velocity (UPV) was evaluated on samples as specified in Standard NF P 18-418 (Afnor 1989). This non-destructive test is based on the propagation of ultrasonic waves in the tested material. Thus, the time transmission was measured using a UPV tester (Fig. 12.7). The velocity of the pulse through the specimen was evaluated using Eq. (12.1):

$$V = \frac{d}{t} \tag{12.1}$$

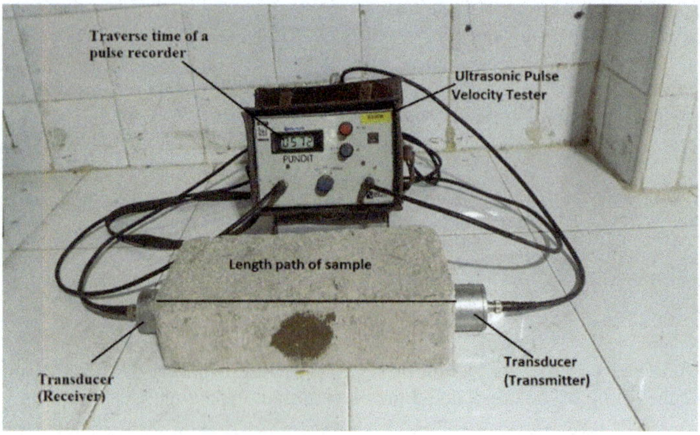

Fig. 12.7 Ultrasonic pulse velocity tester

where V (m/s) is the velocity of the ultrasonic pulse, d (m) is the path length of the sample, and t (s) is the traverse time of the pulse.

The dynamic elastic modulus can be determined using Eq. (12.2):

$$Ed = \rho \times V^2 \tag{12.2}$$

where: Ed (GPa) is the dynamic elastic modulus, and ρ (kg/m^3) is the specimen bulk density.

Compressive Strength Test

Three specimens from each produced mixture of CEB were tested for compressive strength. In this test, the hardened block was subjected to a constant, non-jerky compressive stress until rupture (Fig. 12.8). With a maximum load capacity of 3000 KN, a Hydraulic IGM—France machine was used to record the highest load supported by the sample during the test. The sample was tested at a pressure rise of 0.15 MPa/s until it completely broke. Following that, every block's compressive strength was determined using Eq. (12.3):

$$\sigma c = \frac{Fc}{S} \tag{12.3}$$

where σc (MPa), Fc (N), and S (mm^2) are the compressive strength, the maximum failure force and the contact surface between the CEB and plates used in the test, respectively.

Fig. 12.8 Compressive strength test

Erosion Resistance

The erosion resistance was evaluated on the samples to check the effect of water on the CEBs according to the New Zealand Standard NZS 4298 as shown in Fig. 12.9 (New Zealand Standard 1998). This test consisted of allowing 100 mL of water to drop continuously onto the exposed surface of each block which was inclined 30° from the horizontal axis, from an elevation of 40 cm. The time taken for 100 mL to drip from the container was 40 min (adapted from the mentioned standard of between 20 and 60 min), and the pit depth was measured. The erodibility index was concluded from its value according to the aforementioned standard, and the CEBs were suitable if the pit depth was less than 5 mm. With a pit depth between 5 and 10 mm, the material was considered erosive and corresponded to class 3. At pit depth between 10 and 15 mm, the erosion was assigned to class 4 and the block was considered as very erosive. Finally, for a pit depth of more than 15 mm, the erosion was assigned to class 5 and failed the test.

Fig. 12.9 The erosion resistance test

Results and Discussion

Apparent Density

The density, defined as mass per unit volume (ρ = m/V), is a fundamental property of building materials that governs their structural performance, thermal behavior, and acoustic characteristics and their suitability for building applications. In this work, the apparent density of the blocks was evaluated. The evolution of apparent density as function Alfa fibers for different cement content is depicted in Fig. 12.10 it was noticed that these values varied between 1525 and 1828 kg/m^3. As anticipated, there was a decrease in apparent density when Alfa fibers were incorporated into the formulation of CEB composite. Also, it could be seen that increasing fibers content caused the apparent density to drop. As result, the apparent density of Alfa plant induced a diminution in the CEB apparent density. It was found for 5% of cement that the apparent density value dropped from 1724 kg/m^3 (0% fibers) to 1525 kg/m^3 (0.4% fibers), which correspond to approximately 13%. As can be concluded, the apparent density decrease is small, which can be attributed to the high pressure loading during the preparation of sample (Taallah and Guettala 2016).Consequently, it can be concluded that the first benefit of integrating Alfa fibers into CEB composite is the production of lightweight blocks. This behavior has been highlighted by previous studies (Garrouri et al. 2022; Omrani et al. 2020; Sadouri et al. 2024; Bouchefra et al. 2022; Khoudja et al. 2021). In contrast, it can be seen from Fig. 12.10 that an increase of cement content from 5 to 10% leads to an increase in the apparent density average

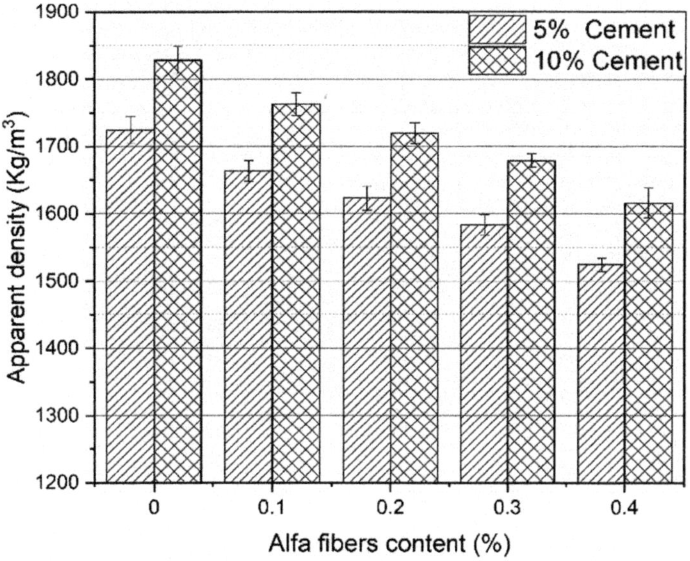

Fig. 12.10 Apparent density of CEB samples at different fiber replacement levels

value of about 6%. This demonstrates that an increase in cement content causes an increase in the CEB apparent density.

Ultrasonic Pulse Velocity

The ultrasonic pulse velocity (UPV) test was conducted to assess the porosity of the CEB and its capacity to absorb sound. Table 12.4 shows the effect of fiber incorporation on the UPV. It was noticed that UPV decreased significantly when the Alfa fibers content in the composite increased. It is important to note that the lowest value was observed for the sample with a content of 0.4% fibers and 5% cement stabilization, which exhibits a reduction of about 44% compared to the non-filled sample with 10% cement stabilization. The variation between the highest and lowest values of UPV is about 37% and 26% for cement contents of 5% and 10% respectively. These findings indicate that the composite filled with Alfa fibers has the ability to decrease the speed of sonic waves' propagation. Therefore, the addition of fibers improved the capacity of CEB to absorb ultrasonic waves and promote the damping vibration, which allows us to obtain a high level of sound insulation. These results could be attributed to the increase in the composite's porosity by Alfa fibers, since the speed of wave propagation in solids is higher than in voids, and also the low UPV of Alfa plants. Similar findings have been seen in the literature (Garrouri et al. 2022; Sadouri et al. 2024; Khoudja et al. 2021; Aamr-Daya et al. 2008; Belakroum et al. 2017). These works demonstrated that construction material composites including vegetable

Table 12.4 Results of density and ultrasonic wave velocity

Cement content (%)	Fibers content (%)	Density (kg/m^3)	Ultrasonic wave velocity (m/s)	Elastic dynamic modulus (Ed) (MPa)
5	0.00	1724 ± 20	1607 ± 24	4.45 ± 0.13
	0.10	1663 ± 15	1497 ± 16	3.72 ± 0.12
	0.20	1623 ± 18	1350 ± 24	2.95 ± 0.20
	0.30	1584 ± 15	1272 ± 20	2.56 ± 0.10
	0.40	1525 ± 10	1167 ± 16	2.07 ± 0.15
10	0.00	1828 ± 20	1690 ± 20	5.22 ± 0.25
	0.10	1763 ± 17	1590 ± 10	4.45 ± 0.14
	0.20	1720 ± 15	1500 ± 18	3.87 ± 0.10
	0.30	1679 ± 10	1420 ± 20	3.38 ± 0.30
	0.40	1616 ± 22	1339 ± 20	2.89 ± 0.10

fibers have good sound absorption characteristics. Table 12.4 also shows the influence of cement content on UPV. It can be noticed from the results that the increase in the cement content from 5 to 10% increased the average value of UPV by about 9%. Based on these results, the increase in UPV could be explained by the increasing amount of calcium silicate (C2S and C3S) created by the increasing cement content. The increasing quantity of these hydrates, which originate from cement hydration, reduces the void network in the matrix and improves its compactness by assuring good bonds between fibers and soil particles, in turn reducing the porosity of the earth block, providing the highest value of UPV.

Dynamic Elasticity Modulus

Table 12.4 summarizes also the influence of Alfa fibers and cement content on the ultrasonic pulse velocity (UPV), it can be noted for 5% of cement that the value varied from 4.45 GPa for blocks without fibers to 2.07 GPa for composite with 0.4% wt fibers. Therefore, the introduction of fibers improved the capacity of blocks to absorb ultrasonic waves and promote the damping vibration, which allows to obtain high level of sound insulation. This finding could be attributed to the increase of the composite porosity by adding vegetal fibers and to the low UPV of Alfa plant. The same tendency has been noticed in the literature (Garrouri et al. 2022; Sadouri et al. 2024; Aamr-Daya et al. 2008). Also, it can be noted that the same behavior has been confirmed for 10% of cement stabilization. On the other side, and based on the obtained results for all fibers addition rates it can be noticed an increase in elasticity dynamic modulus average value from 3.15 GPa (5% Cement) to 3.96 GPa (10% Cement), which correspond to approximately 25%. These results could be due to the increase of block density and decrease of its porosity by increasing the stabilizer percentage, which contributes enhance the compactness of the CEB and to accelerate the speed of wave propagation through the composite material.

Compressive Strength

In this work, the compressive strength of compressed earth blocks filled with Alfa fibers stabilized with cement after 28 days of curing in a laboratory setting was investigated in both dry and wet states. This is because the compressive strength of blocks is considered to be an important indicator of masonry strength. The following sections summarize the obtained results of compressive strength tests performed on the block in dry conditions.

The variation of dry compressive strength (DCS) as a function of fiber content for different cement content is depicted in Fig. 12.11. From the results, it can be seen that for each cement content, the incorporation of fibers leads to a decrease in DCS for all the specimens tested, compared to the composite without fibers, and the highest value corresponds to the sample with a cement content of 10% and without fibers, whereas the lowest value corresponds to the sample with a cement content of 5% and a fiber percentage of 0.4%. The graph shows clearly that an increase in the fiber ratio drops the compressive strength. The decrease in strength between CEB with 0.4% fiber reinforcement and CEB without fibers is 24% for 5% of cement and 16% for 10% of cement. Consequently, the incorporation of fibers has a negative impact on DCS. Figure 12.11 shows also that the DCS increases with an increasing cement content, e.g., the average value of DCS for the unreinforced blocks is 8.16 MPa for 5% of cement and 10.11 MPa for 10% of cement. The obtained results are in good agreement with previous work from the literature on the influence of cement content on the strength of CEBs filled with date palm fibers. Similar finding was noted by Taallah et al. (2014), where same correlations between cement content and DCS were highlighted. They reported in their study that the DCS reached average values of 7.5 MPa, 9.75 MPa, and 12.2 MPa for 5%, 6,5%, and 8% cement contents, respectively. This finding is also in agreement with the finding of Bahar et al. They concluded that the increase in cement from 4 to 12% led to an increase in DCS from 2.2 to 5.00 MPa and that the better result of 6.20 MPa corresponded to 20% cement stabilization for the unreinforced CEBs (Bahar et al. 2004). On the other side, and based on our findings, the decrease in compressive strength for fiber addition could be explained by the low stiffness of Alfa fibers, compared to earth–cement materials. This behavior has been highlighted by previous works (Khedari et al. 2005; MacVicar et al. 1999; Omrani et al. 2020). It was noticed that the incorporation of fibers increases the void network in the blocks, which is the main factor leading to decreased DCS. Moreover, the weak interfacial bond between fibers and the matrix decreases the mechanical performance. Also, the heterogeneous dispersion of fibers during the preparation of CEBs mixture related to their baling effects leads to lower compressive strength. It should be mentioned that the fractures are initiated around the surface of the fibers when the CEB is loaded, which highly accelerates the failure

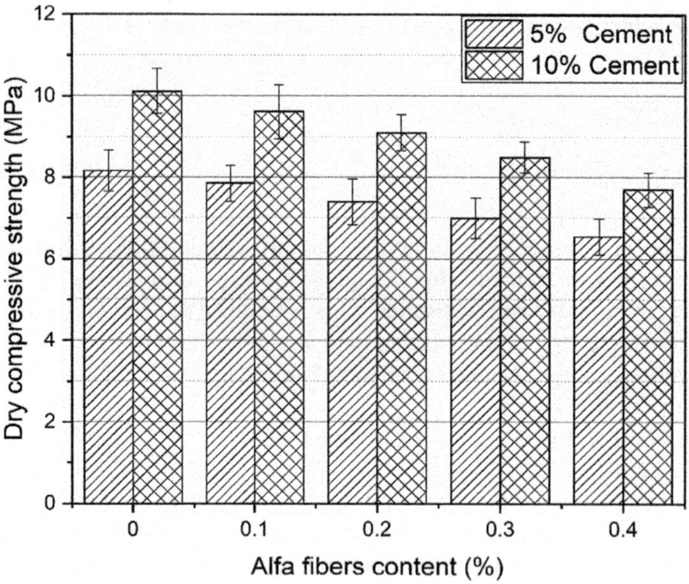

Fig. 12.11 Result of dry compressive strength according to the fiber content

of the composite. Overall, it must be concluded that, in spite of this reduction in strength, the values found for DCS were very satisfactory, since it is recommended that the dry strength for CEBs be superior to 2.4 MPa at 28 days according to Houben et al. (1996).

Erosion Resistance

Following New Zealand Standard NZS 4298, the resistance of earth blocks to water erosion was also used to evaluate their durability (New Zealand Standard 1998). It must be mentioned that, in accordance with the aforementioned standard, resistance to erosion decreased as erosion depth increased. Figure 12.12 shows the erosion depth of the three samples tested for each formulation. As can be seen in this figure, For all samples, the water did not cause any surface deterioration, indicating how effective this CEB stabilization technique is at preventing water erosion. All the tested CEB had an erosive depth inferior to 5 mm, which corresponds to an erodibility index lower than class 3, allowing us to classify the blocks as slightly erosive materials, and thus satisfying the requirement of NZS (New Zealand Standard 1998).

This finding also demonstrates how cement stabilization of earth blocks improves their resistance to erosion. Additionally, the better the erosion resistance, the higher the cement content. The good performance of this composite material in terms of erosion resistance appears to be mainly attributable to the good compactness and

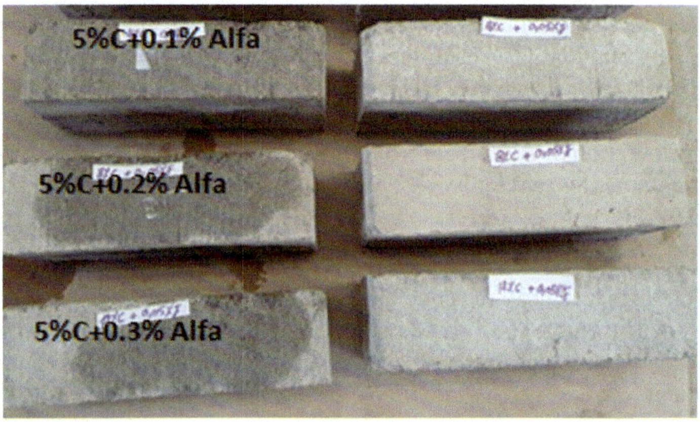

Fig. 12.12 CEBs after being tested for erosion resistance

cohesion between the cement and soil with a low content of fibers, based on visual observations throughout this experimental test. By binding soil particles to fibers and preventing water infiltration into the block, the cement's hydration reaction with the matrix helps to reduce the void network within the mixture. The results of the erosion test demonstrate that the composite material produced during this study can meet the specifications for this type of building material. This is extremely important because, due to their use in masonry and the external loading they endure throughout their service life, CEBs' durability is just as significant as their mechanical strength. The result indicates that CEBs with lower cement and higher fiber content have good erosion resistance which is crucial in areas prone to extreme weather events like floods and storms, which are becoming more frequent and intense due to climate change.

Conclusions

In this study, we investigated the effect of Alfa fiber content and Portland cement stabilization on the physical, mechanical and durability properties of compressed earth blocks. According to the experimental tests, the following conclusions could be drawn:

- The apparent density of the CEB containing 0.40% Alfa fibers was reduced by approximately 20% in comparison to the CEB composite without fibers, regardless of the Portland cement percentage.
- The impact of Alfa fibers on the ultrasonic pulse velocity was favorable to promote sound insulation characteristics and damping vibration, owing to the reduction in its value by up to 44% for 5% cement stabilization reinforced with 0.40% fiber content.

- The integration of fibers into CEB composite reduce the value of elastic dynamic modulus by about 150% regardless the cement content, which is beneficial for vibration dampening, energy absorption and greater bending and deformation before damage.
- The incorporation of Alfa fibers from 0 to 0.4% results in a moderate decrease in the dry compressive strength of CEBs.
- The durability criteria were respected based on the results of the experimental resistance to erosion tests.

These findings lead us to the conclusion that good properties can be obtained in a CEB composite, which can provide safe, comfortable, and long-lasting building materials, by combining Portland cement as a stabilizer with Alfa fibers as a reinforcement.

Acknowledgements The authors gratefully acknowledge Cadi Ayyad University for their support.

Author Contributions R. S.: Conceptualization, Investigation, Visualisation, Methodology, Testing, Writing original draft, Writing—Review and editing. H. K.: Conceptualization, Supervision, Validation, Writing—Review and Editing. M. B.: Conceptualization, Supervision, Validation, Writing—Review and Editing. All authors have read and agreed to the published version of the manuscript.

Funding This research received no external funding. The authors gratefully acknowledge Cadi Ayyad University for their support.

Institutional Review Board Statement Not applicable.

Informed Consent Statement Not applicable.

Data Availability Statement The data presented in this study are available on request from the corresponding author. The data are not publicly available due to privacy.

Conflicts of Interest The authors declare no conflict of interest.

References

Aamr-Daya E, Langlet T, Benazzouk A, Quéneudec M (2008) Feasibility study of lightweight cement composite containing flax by-product particles: physico-mechanical properties. Cem Concr Compos 30(10):957–963

Abhilash HN, Walker P, Venkatarama Reddy B, Heath A, Maskell D (2020) Compressive strength of novel alkali-activated stabilized earth materials incorporating solid wastes. J Mater Civ Eng 32(6):04020118

Achenza M, Fenu L (2006) On earth stabilization with natural polymers for earth masonry construction. Mater Struct 39(1):21–27

Adam E, Jones P (1995) Thermophysical properties of stabilised soil building blocks. Build Environ 30(2):245–253

Afnor N (1989) P18-418, Béton-Auscultation sonique. Mesure du temps de propagation d'ondes soniques dans le béton. Normalisation française, pp 1–12

Afnor X (2001) P13-901: blocs de terre comprimée pour murs et cloisons, definitions-specifications-methodes d'essais-conditions de réception. St.-Denis Plaine Cedex Afnor

Ajouguim S, Abdelouahdi K, Waqif M, Stefanidou M, Saâdi L (2019) Modifications of Alfa fibers by alkali and hydrothermal treatment. Cellulose 26:1503–1516

Alam I, Naseer A, Shah A (2015) Economical stabilization of clay for earth buildings construction in rainy and flood prone areas. Constr Build Mater 77:154–159

Arrigoni A, Beckett C, Ciancio D, Dotelli G (2017) Life cycle analysis of environmental impact vs. durability of stabilised rammed earth. Constr Build Mater 142:128–136

ASTM C 1557-03 (2003) Stand test method tensile strength young's modul fibers. ASTM International

ASTM Committee D-18 on Soil and Rock (2007) Standard test methods for laboratory compaction characteristics of soil using standard effort (12400 ft-lbf/ft^3 (600 kN-m/m^3))[1]. ASTM International

ASTM D (2007) Standard test method for particle-size analysis of soils

ASTM D 4-0 (2010) Standard test methods for liquid limit, plastic limit, and plasticity index of soils. D4318-10

Bahar R, Benazzoug M, Kenai S (2004) Performance of compacted cement-stabilised soil. Cem Concr Compos 26(7):811–820

Belakroum R et al (2017) Hygric buffer and acoustic absorption of new building insulation materials based on date palm fibers. J Build Eng 12:132–139

Binici H, Aksogan O, Bodur MN, Akca E, Kapur S (2007) Thermal isolation and mechanical properties of fibre reinforced mud bricks as wall materials. Constr Build Mater 21(4):901–906

Bouchefra I, Bichri FZE, Chehouani H, Benhamou B (2022) Mechanical and thermophysical properties of compressed earth brick rienforced by raw and treated doum fibers. Constr Build Mater 318:126031

Burroughs S (2010) Recommendations for the selection, stabilization, and compaction of soil for rammed earth wall construction. J Green Build 5(1):101–114

Danso H, Martinson DB, Ali M, Williams JB (2015) Physical, mechanical and durability properties of soil building blocks reinforced with natural fibres. Constr Build Mater 101:797–809

Eires R, Camões A, Jalali S (2017) Enhancing water resistance of earthen buildings with quicklime and oil. J Clean Prod 142:3281–3292

Fawzy S, Osman AI, Doran J, Rooney DW (2020) Strategies for mitigation of climate change: a review. Environ Chem Lett 18:2069–2094

Gandia RM, Gomes FC, Corrêa AAR, Rodrigues MC, Mendes RF (2019) Physical, mechanical and thermal behavior of adobe stabilized with glass fiber reinforced polymer waste. Constr Build Mater 222:168–182

Garrouri S, Lakhal W, Benazzouk A, Sediki E (2022) Potential use of Alfa fibers in construction material: physico-mechanical and thermal characterisation of reinforced specimen. Constr Build Mater 342:127787

Houben H, Rigassi V, Garnier P (1996) Blocs de terre comprimée: équipements de production. CRATerre Brux Belg 149

Islam MS, Iwashita K (2010) Earthquake resistance of adobe reinforced by low cost traditional materials. J Nat Disaster Sci 32(1):1–21

Islam MS, Tausif-E-Elahi, Shahriar AR, Nahar K, Hossain TR (2020) Strength and durability characteristics of cement-sand stabilized earth blocks. J Mater Civ Eng 32(5):04020087

Khedari J, Watsanasathaporn P, Hirunlabh J (2005) Development of fibre-based soil–cement block with low thermal conductivity. Cem Concr Compos 27(1):111–116

Khoudja D, Taallah B, Izemmouren O, Aggoun S, Herihiri O, Guettala A (2021) Mechanical and thermophysical properties of raw earth bricks incorporating date palm waste. Constr Build Mater 270:121824. https://doi.org/10.1016/j.conbuildmat.2020.121824

Kriker A, Debicki G, Bali A, Khenfer M, Chabannet M (2005) Mechanical properties of date palm fibres and concrete reinforced with date palm fibres in hot-dry climate. Cem Concr Compos 27(5):554–564

Limam A, Zerizer A, Quenard D, Sallee H, Chenak A (2016) Experimental thermal characterization of bio-based materials (Aleppo Pine wood, cork and their composites) for building insulation. Energy Build 116:89–95

MacVicar R, Matuana LM, Balatinecz JJ (1999) Aging mechanisms in cellulose fiber reinforced cement composites. Cem Concr Compos 21(3):189–196. https://doi.org/10.1016/S0958-9465(98)00050-X

Mateus R, Fernandes J, Teixeira ER (2019) Environmental life cycle analysis of earthen building materials

Miller SA, Moore FC (2020) Climate and health damages from global concrete production. Nat Clim Change 10(5):439–443

Miller SA, Horvath A, Monteiro PJ (2018) Impacts of booming concrete production on water resources worldwide. Nat Sustain 1(1):69–76

Minke G (2013) Building with earth. In: Building with earth. Birkhäuser

Mostafa M, Uddin N (2016) Experimental analysis of compressed earth block (CEB) with banana fibers resisting flexural and compression forces. Case Stud Constr Mater 5:53–63

Nagaraj H, Sravan M, Arun T, Jagadish K (2014) Role of lime with cement in long-term strength of compressed stabilized earth blocks. Int J Sustain Built Environ 3(1):54–61

Ness DA, Xing K (2017) Toward a resource-efficient built environment: a literature review and conceptual model. J Ind Ecol 21(3):572–592

New Zealand Standard (1998) NZS 4298: materials and workmanship for earth buildings. Wellington

Omrani H, Hassini L, Benazzouk A, Beji H, ELCafsi A (2020) Elaboration and characterization of clay-sand composite based on Juncus acutus fibers. Constr Build Mater 238:117712–118712

Panesar D, Shindman B (2012) The mechanical, transport and thermal properties of mortar and concrete containing waste cork. Cem Concr Compos 34(9):982–992

Sadouri R, Kebir H, Benyoucef M (2024) The effect of incorporating Juncus fibers on the properties of compressed earth blocks stabilized with Portland cement. Appl Sci 14(2):815

Saghrouni Z, Baillis D, Jemni A (2020) Composites based on Juncus maritimus fibers for building insulation. Cem Concr Compos 106:103474

Segetin M, Jayaraman K, Xu X (2007) Harakeke reinforcement of soil–cement building materials: manufacturability and properties. Build Environ 42(8):3066–3079

Sujatha ER, Devi SS (2018) Reinforced soil blocks: viable option for low cost building units. Constr Build Mater 189:1124–1133

Taallah B, Guettala A (2016) The mechanical and physical properties of compressed earth block stabilized with lime and filled with untreated and alkali-treated date palm fibers. Constr Build Mater 104:52–62

Taallah B, Guettala A, Guettala S, Kriker A (2014) Mechanical properties and hygroscopicity behavior of compressed earth block filled by date palm fibers. Constr Build Mater 59:161–168. https://doi.org/10.1016/j.conbuildmat.2014.02.058

Williams J, Lawrence M, Walker P (2016) A method for the assessment of the internal structure of bio-aggregate concretes. Constr Build Mater 116:45–51

Zak P, Ashour T, Korjenic A, Korjenic S, Wu W (2016) The influence of natural reinforcement fibers, gypsum and cement on compressive strength of earth bricks materials. Constr Build Mater 106:179–188

Zare P, Narani SS, Abbaspour M, Fahimifar A, Hosseini SMMM, Zare P (2020) Experimental investigation of non-stabilized and cement-stabilized rammed earth reinforcement by waste tire textile fibers (WTTFs). Constr Build Mater 260:120432

Zhang L, Gustavsen A, Jelle BP, Yang L, Gao T, Wang Y (2017) Thermal conductivity of cement stabilized earth blocks. Constr Build Mater 151:504–511

Correction to: Exploring Climate Change: Morocco in Focus

Kholoud Kahime, Abdelali Boussakouran, Moulay Abdelmonaim El Hidan, and Mohamed El Yamani

Correction to:
Chapter 1 in: K. Kahime et al. (eds.), *Climate Change Effects and Sustainability Needs*, Springer Climate, https://doi.org/10.1007/978-3-031-59603-2_1

The original version of the book was inadvertently published with incorrect corresponding author identification and the sequence of the authors in Chapter 1, which have now been corrected. The book and the chapter have been updated with the changes.

The updated version of this chapter can be found at
https://doi.org/10.1007/978-3-031-59603-2_1

K. Kahime et al. (eds.), *Climate Change Effects and Sustainability Needs*, Springer Climate, https://doi.org/10.1007/978-3-031-59603-2_13

Index

K. Kahime et al. (eds.), *Climate Change Effects and Sustainability Needs*, Springer Climate, https://doi.org/10.1007/978-3-031-59603-2

SPRINGER NATURE

GPSR Compliance

The European Union's (EU) General Product Safety Regulation (GPSR) is a set of rules that requires consumer products to be safe and our obligations to ensure this.

If you have any concerns about our products, you can contact us on ProductSafety@springernature.com

In case Publisher is established outside the EU, the EU authorized representative is:

Springer Nature Customer Service Center GmbH
Europaplatz 3
69115 Heidelberg, Germany

The manufacturer's authorised representative in the EU is Springer Nature Customer Service Centre GmbH, Europaplatz 3, 69115 Heidelberg, Germany. If you have any concerns regarding our products, please contact ProductSafety@springernature.com

Printed and bound by CPI Group (UK) Ltd, Croydon, CR0 4YY
01/10/2026
02225907-0001